# Down Syndrome

## Advances in Medical Care

# Down Syndrome

## Advances in Medical Care

### Editors

**Ira T. Lott**

Departments of Pediatrics and Neurology
University of California Irvine
Irvine, California

**Ernest E. McCoy**

Department of Pediatrics
University of Alberta
Edmonton, Alberta, Canada

# WILEY-LISS

A JOHN WILEY & SONS, INC., PUBLICATION

New York • Chichester • Brisbane • Toronto • Singapore

**Address All Inquiries to the Publisher**
**Wiley-Liss, Inc., 605 Third Avenue, New York, NY 10158-0012**

**Copyright © 1992 Wiley-Liss, Inc.**

**Printed in the United States of America.**

**While the authors, editors, and publisher believe that drug selection and dosage and the specification and usage of equipment and devices, as set forth in this book, are in accord with current recommendations and practice at the time of publication, they accept no legal responsibility for any errors or omissions, and make no warranty, express or implied, with respect to material contained herein. In view of ongoing research, equipment modifications, changes in governmental regulations and the constant flow of information relating to drug therapy, drug reactions, and the use of equipment and devices, the reader is urged to review and evaluate the information provided in the package insert or instructions for each drug, piece of equipment, or device for, among other things, any changes in the instructions or indication of dosage or usage and for added warnings and precautions.**

**Second Printing, September 1992**
**Third Printing, July 1993**

**Library of Congress Cataloging-in-Publication Data**
Down Syndrome: advances in medical care / editors, Ira T. Lott,
   Ernest E. McCoy
      p.  cm.
      Proceedings of a Health Care Conference held under the auspices
   of the National Down Syndrome Society, in San Diego, Calif., March
   15-16, 1991.
      Includes bibliographical references and index.
      ISBN 0-471-56181-9 casebound
      ISBN 0-471-56184-3 paperbound
      1. Down's Syndrome--Congresses. I. Lott, Ira T. II. McCoy,
   Ernest E. III. National Down Syndrome Society (U.S.) IV. Health
   Care Conference (1991 : San Diego, Calif.)
      [DNLM: 1. Down's Syndrome--complications--congresses. 2. Down's
   Syndrome--genetics--congresses. 3. Down's Syndrome--therapy-
   -congresses. WS 107 D7487 1991]
   RJ506.D68D673 1992
   616.85 ' 8842--dc20
   DNLM/DLC
   for Library of Congress                                    91-40009
                                                              CIP

**The text of this book is printed on acid free paper.**

# Contents

Contributors . . . . . . . . . . . . . . . . . . . . . . . . . . . . . . . . . . . vii

Foreword
Charles J. Epstein . . . . . . . . . . . . . . . . . . . . . . . . . . . . . . . . ix

Preface
Ira T. Lott and Ernest E. McCoy . . . . . . . . . . . . . . . . . . . . . . . . xi

Acknowledgments . . . . . . . . . . . . . . . . . . . . . . . . . . . . . . . . xiii

Timetable . . . . . . . . . . . . . . . . . . . . . . . . . . . . . . . . . . . . . xv

PRINCIPLES OF DEVELOPMENT

Advances in the Understanding of Chromosome 21 and Down Syndrome
Julie R. Korenberg, Stefan-M. Pulst, and Scott Gerwehr . . . . . . . . . . . . . . . 3

Current and Future Modes of Prenatal Diagnosis
Mark H. Bogart . . . . . . . . . . . . . . . . . . . . . . . . . . . . . . . . . 13

Analysis of Non-Disjunction in Human Autosomal Trisomies
Terry Hassold, Sallie Freeman, Carol Phillips, Stephanie Sherman, and Norma Takaesu . 23

LEARNING AND COGNITION

Learning and Cognition in Down Syndrome
Lynn Nadel . . . . . . . . . . . . . . . . . . . . . . . . . . . . . . . . . . . 37

Development of Speech and Language in Children With Down Syndrome
Jon F. Miller . . . . . . . . . . . . . . . . . . . . . . . . . . . . . . . . . . 39

CLINICAL ADVANCES

The Person With Down Syndrome: Medical Concerns and Educational
Strategies
Siegfried M. Pueschel . . . . . . . . . . . . . . . . . . . . . . . . . . . . . . 53

v

Cardiorespiratory Problems in Children With Down Syndrome
Langford Kidd . . . . . . . . . . . . . . . . . . . . . . . . . . . . . . . . . . . . **61**

Endocrine Function in Down Syndrome
Ernest E. McCoy . . . . . . . . . . . . . . . . . . . . . . . . . . . . . . . . . **71**

Susceptibility to Infectious Disease in Down Syndrome
David J. Lang . . . . . . . . . . . . . . . . . . . . . . . . . . . . . . . . . . . **83**

Hematologic and Oncologic Disorders in Down Syndrome
Alvin Zipursky, Annette Poon, and John Doyle . . . . . . . . . . . . . . . . . . . **93**

Neurological and Neurobehavioral Disorders in Down Syndrome
Ira T. Lott . . . . . . . . . . . . . . . . . . . . . . . . . . . . . . . . . . . . 103

Orthopedic Disorders in Down Syndrome
Liebe S. Diamond . . . . . . . . . . . . . . . . . . . . . . . . . . . . . . . . 111

Recurrent Otitis and Sleep Obstruction in Down Syndrome
Marshall Strome and Scott Strome . . . . . . . . . . . . . . . . . . . . . . . . 127

Oral and Dental Considerations in Down Syndrome
Edward S. Sterling . . . . . . . . . . . . . . . . . . . . . . . . . . . . . . . 135

Ocular Disorders in Down Syndrome
Thomas D. France . . . . . . . . . . . . . . . . . . . . . . . . . . . . . . . 147

**ADULTS WITH DOWN SYNDROME**

Sexuality and Community Living
William E. Schwab . . . . . . . . . . . . . . . . . . . . . . . . . . . . . . . 157

Aging and Alzheimer's Disease in People With Down Syndrome
Krystyna E. Wisniewski, A. Lewis Hill, and Henry M. Wisniewski . . . . . . . . . 167

Index . . . . . . . . . . . . . . . . . . . . . . . . . . . . . . . . . . . . . . 185

# Contributors

**Mark H. Bogart,** Division of Medical Genetics, Department of Medicine, University of California, San Diego, La Jolla, CA 92093-0639 [13]

**Liebe S. Diamond,** Department of Orthopaedics, University of Maryland, Baltimore, MD 21201 [111]

**John Doyle,** Department of Pediatrics, Division of Hematology/Oncology, The Hospital for Sick Children and The University of Toronto, Toronto, Ontario M5G 1X8, Canada [93]

**Charles J. Epstein,** Department of Pediatrics and Department Biochemistry and Biophysics, University of California, San Francisco, CA 94143 [ix]

**Thomas D. France,** Department of Ophthalmology, University of Wisconsin Medical School, Madison, WI 53792 [147]

**Sallie Freeman,** Department of Pediatrics, Division of Medical Genetics, Emory University School of Medicine, Atlanta, GA 30322 [23]

**Scott Gerwehr,** Division of Genetics, Ahmanson Department of Pediatrics, Cedars-Sinai Medical Center, University of California, Los Angeles, CA 90048 [3]

**Terry Hassold,** Department of Pediatrics, Division of Medical Genetics, Emory University School of Medicine, Atlanta, GA 30322 [23]

**A. Lewis Hill,** Department of Pathological Neurobiology, New York State Institute for Basic Research in Developmental Disabilities, Staten Island, NY 10314 [167]

**Langford Kidd,** Division of Pediatric Cardiology, The Johns Hopkins University School of Medicine, Baltimore, MD 21205 [61]

**Julie R. Korenberg,** Division of Genetics, Department of Pediatrics, Cedars-Sinai Medical Center, University of California, Los Angeles, CA 90048 [3]

**David J. Lang,** Pediatrician in Chief, Director of Medical Education, Children's Hospital of Orange County, Orange, CA 92668 [83]

**Ira T. Lott,** Departments of Pediatrics and Neurology, University of California Irvine, Irvine, CA 92668 [xi, 103]

**Ernest E. McCoy,** Department of Pediatrics, Walter MacKenzie Health Sciences Center, University of Alberta, Edmonton, Alberta T6G 2R7, Canada [xi, 71]

**Jon F. Miller,** Department of Communicative Disorders, Waisman Mental Retardation Research Center, University of Wisconsin-Madison, Madison, WI 53706 [39]

**Lynn Nadel,** Department of Psychology, University of Arizona, Tucson, AZ 85721 [37]

The number in brackets is the opening page number of the contributor's article.

**Carol Phillips,** Department of Pediatrics, Division of Medical Genetics, Emory University School of Medicine, Atlanta, GA 30322 [23]

**Annette Poon,** Department of Pediatrics, Division of Hematology/Oncology, The Hospital for Sick Children and The University of Toronto, Toronto, Ontario M5G 1X8, Canada [93]

**Siegfried M. Pueschel,** Department of Pediatrics, Child Development Center, Brown University Program in Medicine, Rhode Island Hospital, Providence, RI 02903 [53]

**Stefan-M. Pulst,** Department of Medicine, Cedars-Sinai Medical Center, University of California, Los Angeles, CA 90048 [3]

**William E. Schwab,** Department of Family Medicine and Practice, University of Wisconsin-Madison, Madison, WI 53706 [157]

**Stephanie Sherman,** Department of Pediatrics, Division of Medical Genetics, Emory University School of Medicine, Atlanta, GA 30322 [23]

**Edward S. Sterling,** Division of Pediatric Dentistry, Nisonger Center of Mental Retardation and Developmental Disabilities, Ohio State University College of Dentistry, Columbus, OH 43210 [135]

**Marshall Strome,** Department of Otolaryngology, Harvard University Medical School, Beth Israel Hospital, Boston, MA 02134 [127]

**Scott Strome,** Department of Otolaryngology, Harvard University Medical School, Beth Israel Hospital, Boston, MA 02134 [127]

**Norma Takaesu,** Department of Pediatrics, Division of Medical Genetics, Emory University School of Medicine, Atlanta, GA 30322 [23]

**Henry M. Wisniewski,** Department of Pathological Neurobiology, New York State Institute for Basic Research in Developmental Disabilities, Staten Island, NY 10314 [167]

**Krystyna E. Wisniewski,** Department of Pathological Neurobiology, New York State Institute for Basic Research in Developmental Disabilities, Staten Island, NY 10314 [167]

**Alvin Zipursky,** Division of Hematology/Oncology, The Hospital for Sick Children, Toronto, Ontario M5G 1X8, Canada [93]

# Foreword

This volume and the conference that served as its genesis spring from an interest in the health of persons with Down syndrome. As the following chapters reveal, the advances made during the past two decades in the care of people with Down syndrome have been truly impressive. However, it is only fair to point out that these improvements in care are much more the result of social changes than they are of changes in medicine and medical practice, as important as the latter may be. What have these social changes been?

**Deinstitutionalization.** Although the adoption of policies designed to restrict the entry of infants and children with Down syndrome into state institutions for the mentally retarded was based principally on fiscal concerns, the benefits to the person with Down syndrome have been enormous. Of greatest significance is the effect on intellectual development, and it has been estimated that home placement, as opposed to institutional placement, has increased the measured IQ by 10 to 20 points. Similarly, although not specifically documented, it is not unlikely that removal from institutions of persons with Down syndrome, with their increased susceptibility to infection, has led to an increase in their life spans.

**Acceptance.** Coupled with deinstitutionalization and the realization that persons with Down syndrome can be cared for within the home and community has been their acceptance within the community. The situation of persons with Down syndrome has been further enhanced by the general increase in public awareness of the strengths and rights of individuals with a large variety of developmental disabilities.

**Access to medical care.** One of the consequences of acceptance has been an improvement in the access of persons with Down syndrome to medical care and an increased willingness of physicians and other health care providers to approach people with Down syndrome in the same manner as persons without this genetic disorder. While this increased access has attended many different aspects of care, it has had a particular effect on the management of infants with Down syndrome who have serious, but surgically correctable, congenital malformations.

**Attention to special needs and vulnerabilities.** Growing out of all of the factors just cited, but also dependent upon increased knowledge of the medical and biological aspects of Down syndrome, has been an increased attention to the special needs and vulnerabilities of the infants and children and, more and more in recent years, adults with Down syndrome. Such attention will be particularly necessary as the life span of persons with Down syndrome continues to lengthen.

Despite the uniqueness of the etiology of Down syndrome, there is really nothing unique about either the therapies and management strategies that are being employed or the means for preventing Down syndrome that are currently available to us. Stated another way, although we know that Down syndrome results from the presence of all or part of a third chromosome 21 in the genome, we have not been able to capitalize on *this* knowledge either to prevent Down syndrome from occurring or to prevent or treat many of the components of the syndrome. Among the latter are, in particular, the developmental (mental) retardation, the increased frequency of leukemia, the greater susceptibility to infection, and—perhaps of greatest concern in the long run—the development of Alzheimer disease.

The reasons for our failure to capitalize on our genetic and molecular knowledge are clear enough. We just do not know enough—enough about why nondisjunction increases with maternal age, about what the genes on chromosome 21 are doing, and about what happens when we have extra copies of these genes. But, the situation is certainly not bleak at all, and things are changing rapidly: the human genome is being mapped, and chromosome 21, in particular, is being mapped and dissected, and its genes are being identified; animal models to study the pathogenesis of Down syndrome are being developed.

The recent major advances in molecular biology, cell biology, and genetics now make it possible for us to approach the problems of Down syndrome in entirely new ways, and the same holds true for advances in the neural and behavioral sciences as well. Things have never been more promising for developing an understanding of the causation and pathogenesis of Down syndrome and for using that understanding to enhance our ability to improve the physical and intellectual status of persons with this condition.

**Charles J. Epstein**

# Preface

This book had its genesis in a Health Care Conference held under the auspices of the National Down Syndrome Society. The purpose of this conference was to bring together professionals who had worked and published in areas relevant to the medical care of individuals with Down syndrome across the life span and to relate advances in the basic sciences to clinical treatment. To our knowledge, this is the first national conference of this type to focus on advances in medical care for individuals with Down syndrome.

The introductory section focuses on Principles of Development in Down Syndrome. In a discussion of genetic advances, attention is given to progress in the physical mapping of chromosome 21 which has allowed some correlations with the clinical expression of the disorder. Sophisticated prenatal techniques which have made possible early intrauterine diagnosis of Down syndrome are reviewed. A discussion of the use of DNA polymorphisms to elucidate the parental origin of the additional chromosome is included. Principles of language and cognition in Down syndrome are considered in the second section in the context of diagnostic approaches to early vocabulary development and how these translate into teaching programs.

Clinical advances in Down syndrome are classified by specialty in section three. Highlights include: 1) a clinical schema for the diagnosis and treatment of cardio-pulmonary disorders; 2) the factors accounting for the susceptibility to infectious disease and some preventative suggestions; 3) an update on the understanding of endocrine disorders with particular emphasis on thyroid abnormalities; 4) a clinical review of advances in leukemia and how they bear upon diagnosis and treatment; 5) the import of structural and functional abnormalities of brain upon the cognitive deficit, hypotonia, seizures and neurobehavioral disturbances; 6) a practical approach to orthopedic intervention; 7) early intervention for otolaryngologic disease so as to prevent consequences for intellectual development and susceptibility to infection; 8) the principle indications for plastic surgical intervention; 9) the consequences of oral/dental pathology and treatment strategies and 10) treatable ocular disorders in the syndrome.

In part as the result of the medical and surgical advances detailed above, the child with Down syndrome can look forward to an average life of fifty-five years.

As a result, medical professionals are facing issues of sexuality and community living, which are explored in the final section, "Adults with Down Syndrome." The aging process with specific emphasis on Alzheimer's disease rounds out this section. Herein the enigmatic relationship between brain pathology and clinical dementia is reviewed as it pertains to the adult with Down syndrome.

We attempt to direct the information gathered in this book towards a timetable for medical/surgical intervention across the life span of individuals with Down syndrome. While we expect that any such proposal will change over time with future advances, the timetable offers a starting point for health care planning for the individual with Down syndrome. The best outcome will be the need to revise the intervention timetable because of advances in health care which take us beyond our current understanding of these important issues for all people with Down syndrome.

**Ira T. Lott, MD**
**Ernest E. McCoy, MD**

# Acknowledgments

The first conference on Down Syndrome Health Care was sponsored by the National Down Syndrome Society, under the leadership of Elizabeth Goodwin, President. The organization of the conference was expertly handled by Donna M. Rosenthal, Executive Director, without whose efforts neither the conference nor this volume would have been possible.

The Society is deeply appreciative of the very generous support of The Procter & Gamble Company, maker of Ivory Bar Soap; The Block Drug Company; Lehn & Fink Canada; and Lederle Laboratories. Their contributions have made the conference and this book of proceedings a reality.

The University of California at Irvine Office of Continuing Medical Education provided course credit for this important program.

Additional information about the work of the National Down Syndrome Society can be obtained from the Society at 666 Broadway, New York, New York, 10012. The telephone numbers of the Society are (212) 460-9330 and (800) 221-4602.

# Timetable for Medical-Surgical Intervention in Down Syndrome

The following timetable comprises a list of suggestions for intervening in the medical-surgical care of individuals with Down syndrome across the life span. The suggestions are based upon recommendations made in this monograph as well as existing preventive medical checklists (Down Syndrome, Papers and Abstracts for Professionals, Vol 12, #2, 1989). One cannot be dogmatic in the application of these guidelines as there are genuine differences in opinion in regard to some of the items. Nonetheless, we feel that this timetable may be used as a guide for organizing the medical care of individuals with Down syndrome. The timetable is directed towards those items which are of particular importance in Down syndrome and should be considered as adjuncts to standard medical care at each age epoch. We anticipate that changes in this schema will evolve as new medical insights improve treatment and preventive measures for all individuals with Down syndrome.

## NEONATAL AND EARLY INFANCY

Establish chromosomal karyotype and communicate diagnosis to parents; refer family to parent support groups

Conduct red reflex exam and screen for cataracts
Check for blockage in gastrointestinal track (vomiting or absence of stools)
Examine infant for congenital heart disease (cyanosis, murmur, irregular heart rhythm; EKG and echocardiogram are strongly recommended)
Conduct thyroid function tests (TSH, T4 and T3)
Hearing screen (auditory brain stem evoked responses)
Enroll in early intervention program (home-based or center based)

## PRESCHOOL (1-5 YEARS)

Office orthopedic exam (forefoot anomalies, bunions, dislocated hips); some recommend initial screening for atlanto-axial dislocation at 2 years while others wait until cartilaginous bone is better developed at 5 years

Initial dental exam (2 years) with attention to fluoride supplementation, periodontal therapy and use of sealants

In addition to normal vaccination schedule, consider influenzal, pneumococcal, and hepatitis B vaccines for children at major risk

Microscopic otoscopy by 18 months if ear canals are too small to allow routine evaluation; yearly audiogram

Complete ophthalmological exam at age 12 months

Annual T4, TSH and T3

Developmental evaluation (physical therapy, occupational therapy, feeding evaluation); continue early intervention programming

Monitor speech and language progress from first words including emphasis on eating and tongue behavior

Preventive behavioral checklist (sleep patterns, nutrition, toileting skills, self care, and communication)

## SCHOOL AGE

Obtain cervical spine X-rays every 5 years if initial screen shows subluxation; also obtain cervical spine films before general anesthesia is given because of hyperextended position

Annual dental exam for orthodontic, periodontics and sealants

Annual audiogram

Annual ophthalmological exam

Annual thyroid screen

Psychoeducational evaluation every 3 years

Preventive behavior checklist (school adjustment, aggression, self injurious behavior, property destruction, transitional observations)

Nutritional and dietary counseling and exercise program

## YOUNG ADULT

Annual TSH, T4, T3

Counseling in regard to pre-vocational adjustment, sexuality, separation from parents, and independent living

Twice yearly dental exams

Daily exercise program; dietary counseling

## ADULT

Aging health maintenance (diabetes, cancer screening)

Exercise program

Screen for symptoms of Alzheimer's disease (personality change, loss of daily living skills, changes in gait, seizures)

Symptoms of depression—look for changes in personality and other symptoms overlapping with Alzheimer's disease

Annual thyroid screen

# Principles of Development

# Advances in the Understanding of Chromosome 21 and Down Syndrome

Julie R. Korenberg, Stefan-M. Pulst, and Scott Gerwehr

Down Syndrome (DS) is a major cause of mental retardation and congenital heart disease affecting the welfare of over 300,000 individuals and their families in the USA alone. In addition Down Syndrome is associated with a characteristic set of facial and other physical features, congenital gut disease, an increased risk of leukemia, defects of the immune system, endocrine abnormalities, and an Alzheimer-like dementia (reviewed in Epstein, 1986; Pueschel 1982). Moreover, in addition to being a major public health concern, Down Syndrome is the prototype for the study of human aneuploidy.

The current revolution in human molecular genetics now provides the possibility to understand the basis of the associated defects. This represents a necessary first step leading to the prevention, amelioration and perhaps eventual treatment of some of the devastating defects. This chapter will briefly trace the understanding of Down syndrome from its beginnings as a clinical condition through the discovery of its chromosomal basis and its recent molecular understanding.

## HISTORY

The story of Down Syndrome (references in Zellweger, 1977) begins about 150 years ago with the first complete description of Down Syndrome by Seguin (1846). However, it was the brief report by Down (1866), "Observations on the ethnic classification of idiots", that established the eponym, and misguided the following generations of scientists and physicians by incorrectly comparing the inner epicanthal fold of Down Syndrome to the extension of the tarsal epicanthal fold seen in oriental populations. The first extensive description of Down Syndrome was provided in the report of 62 cases of "Kalmuc Idiocy" by Fraser and Mitchell (1876). This provided a complete physical description, noted the increased risk of Down Syndrome with advanced maternal age and

3

described Down Syndrome neuropathology. The observation of Down Syndrome congenital heart anomalies by Garrod and Thompson (1898) then rounded out the basic clinical description of Down Syndrome, largely complete by the turn of the century. The next major advance came 60 years later, requiring a deeper understanding of the biological basis of human heredity. Although abnormality of chromosome number had been suspected earlier from a knowledge of plant chromosomal variation, it was only in 1959 that LeJeune and Jacobs independently determined that Down Syndrome was caused by trisomy 21. This understanding of Down Syndrome as resulting from the duplication of genes on chromosome 21 was focused by the suggestion of Niebuhr (1974), that the "typical Down Syndrome phenotype" might be caused by the duplication of only a part of chromosome 21 band q22 which itself represents about one half of the long arm. The observation of Down Syndrome in a chimpanzee (McClure, 1969) provided compelling evidence that supported the growing understanding that the specific phenotypic features of Down Syndrome are caused by the duplication of specific genes located on chromosome 21.

Progress in the genetic and physical mapping of chromosome 21 has now reached the point at which it is possible to begin the correlation of the phenotypic components of Down Syndrome with imbalance of specific regions of the chromosome. Several preliminary efforts in this direction have already been made and suggest that the phenotypic and molecular analysis of the relatively rare individuals with chromosome 21 duplications ("partial trisomy") can be used to specify which regions of chromosome 21 are involved in the generation of specific components of the phenotype. The ultimate goal of correlating genotype with phenotype (phenotypic mapping) is to make it possible to discover which particular genes are responsible for which aspects of the phenotype, thereby permitting the pathogenesis of the syndrome to be elucidated and, hopefully, its most serious consequences to be prevented or ameliorated.

## CURRENT STATUS OF GENOTYPE-PHENOTYPE CORRELATIONS IN DOWN SYNDROME

It is now over 30 years since Down Syndrome was found to be caused by trisomy 21 and more than 15 years have elapsed since the role of band q22 in causing the Down Syndrome phenotype was suggested. Two changes in direction have recently been defined. First, it is now clear that genes in other regions contribute significantly to the phenotype. Second, the emergence of the physical map of chromosome 21 has eliminated the uncertainty of *cytogenetic* analyses and has made possible the molecular definition of regions responsible for specific phenotypic features of Down Syndrome. A phenotypic map of Down Syndrome based on cytogenetic analyses is shown in Figure 1. This map, constructed from 17 well-defined cases of chromosome 21 duplications published since 1973, shows the overlaps of duplicated regions which are associated with the feature(s) indicated. The map must, for two reasons, be considered as indicating only the

DOWN SYNDROME PHENOTYPIC MAP
CYTOGENETIC
(1973 - 1989)

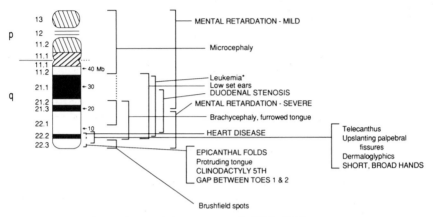

Fig. 1.   Down syndrome phenotypic map: cytogenetic (1973–1989).

minimal regions involved in producing a particular feature. Too little information derives from small duplications, and the data are incomplete with regard to the lack of features in many of the patients. Furthermore, such analyses do not indicate the number of genes involved.

## MOLECULAR STRUCTURE OF CHROMOSOME 21

Knowledge of the physical and genetic maps of chromosome 21 underlies the molecular and cytogenetic methods used to determine DNA sequence copy number in rearranged chromosomes 21. Long-range restriction maps of the long arm of chromosome 21 have recently been constructed using somatic cell hybrids, irradiation reduction hybrids, Southern blot hybridization, and pulsed field gel electrophoresis (Carrit and Litt 1989; Gardiner et al. 1990; Burmeister et al. 1990; Cox et al. 1990).

These recent advances have allowed the creation of a detailed molecular map of chromosome 21 which consists of greater than 75 ordered DNA fragments, or about one per 50-400 kb. These recognize an ordered series of pulsed field fragments and include expressed genes. (Gardiner et al. 1990; Cox et al. 1990; Burmeister et al. 1990). In addition, a genetic linkage map of the chromosome is emerging (Tanzi et al. 1988, Warren et al. 1989). Knowledge of this map provides the opportunity to define, usually within 400 kb, the duplicated regions associated with the features of Down Syndrome. Given the rapid progress of the development of the chromosome 21 molecular map and the observed clustering of spontaneous breakpoints (Gardiner et al. 1990), this precision is likely to increase to the range of 50-100 kb, well within the range needed to define the expressed genes responsible for Down Syndrome.

## DOWN SYNDROME PHENOTYPIC MAP AND CHROMOSOME 21 GENES

The phenotypic map of Down Syndrome based on the molecular analysis of chromosome 21 duplications is now emerging from studies conducted by a number of groups. The technique of Southern blot dosage analysis used by our laboratory to estimate gene copy number, is illustrated in Fig. 2. This shows an autoradiogram of a representative Southern blot used for quantitative densitometric analysis of DNA sequence copy number in patients with partial trisomy 21. Lanes containing DNA from a patient with partial trisomy 21 (DUP21JS) and from diploid control (C) are shown. DNA sequences for SODI (CuZn superoxide dismutase) and *D21S47* are present in three copies, whereas the DNA sequence *D21S44* is present in 2 copies. The DNA sequence designated *D17S33* is the chromosome 17 reference probe. The results of these and other studies indicate a general consensus that although the facial features of Down Syndrome may be determined by genes in the region of D21S55-21qter (Rahmani et al. 1989; McCormick et al. 1989; Korenberg et al. 1990c), the mental retardation results

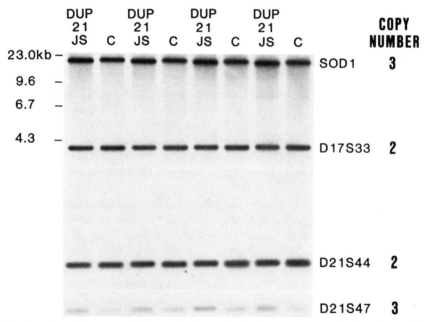

Fig. 2. Autoradiogram used for quantitative analysis of gene copy number in patients with partial trisomy 21.

from imbalance of genes mapping throughout the chromosome. Moreover, the minimal regions likely to contain the gene(s) determining the congenital heart disease and the duodenal stenosis have also been defined as D21S55-21qter and D21S8-D21S15 (Korenberg et al. 1990a, 1990c, 1991b), respectively.

These regions and the genes mapping in them are shown in Figures 3 and 4 and in Table 1. Table 1 also indicates gene sequences that are known to be expressed but whose function is unknown. It is of interest that the region defined for the Down Syndrome congenital heart disease comprises about 9 million base pairs and includes the ETS2 oncogene, the ETS2 related gene (ERG) and the gene for CBS.

However the region does not include SODI nor the genes involved in purine biosynthesis (PRGS, PAIS, PGFT). Similarly, the region defined for the duodenal stenosis excludes all the genes located between MX1/2 and the telomere but still includes APP, SOD1 the genes for purine biosynthesis and ETS, among others. It is important to note that although any of the genes included in a given region may contribute to a feature, the genes now defined in any region likely represent less

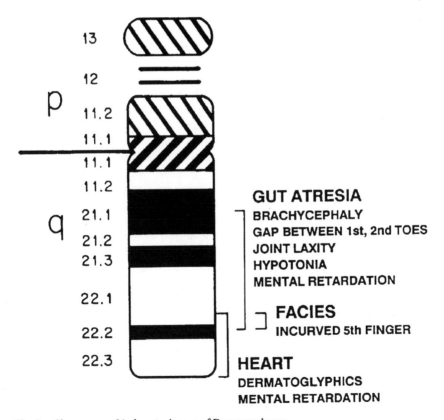

Fig. 3.  Chromosome 21 phenotypic map of Down syndrome.

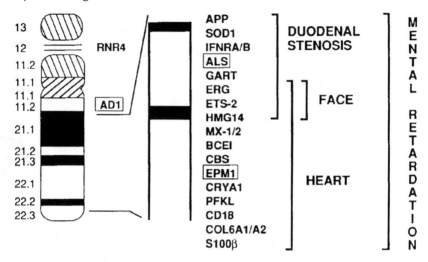

Fig. 4.   Chromosome 21 expressed genes and Down syndrome.

than 10% of all genes located in that region of chromosome 21. Moreover, the regions defined are still large, the features are complex and little is known at the molecular level about the factors that influence their development.

Nonetheless, for other Down Syndrome features, it is tempting to speculate on the potential relationship between gene and phenotype. For example, it is possible that abnormalities in the expression of CRYA1 (the major structural protein of the ocular lens) may predispose to the cataracts seen in Down Syndrome or that the IFN A/BR, the interferon receptors or CD18, may be related to the immune deficits and, finally, that the oncogene ETS2 may be related to one of the Down Syndrome leukemias (Zipursky et al., 1987). Further, the potential relationship of the Down Syndrome Alzheimer-like dementia to the duplication of APP (the amyloid precursor protein) remains an open question. Although our previous results indicate that duplication of APP is not necessary for most Down Syndrome features, clearly it may be responsible for others, including the dementia. In addition to the cloned genes and expressed DNA sequences located on chromosome 21, there are recently additional neurological diseases mapped by family studies. These include Progressive myoclonus epilepsy (Unverricht-Lundborg) (PMEI) (Lehesjoki et al. 1991) and Amyotrophic lateral sclerosis (ALS) (Siddique et al. 1991) in addition to Alzheimer disease (AD).

## CHROMOSOME 21 GENES FOR DOWN SYNDROME MENTAL RETARDATION

Definition of the genes for mental retardation (MR) in Down Syndrome present particular challenges. With the exception of a single report (Daniel, 1979), individuals trisomic for any region of chromosome 21 are mentally retarded. This

**Table 1.    Genes and Expressed Sequences on Chromosome 21**

| DNAs | GENES | |
|---|---|---|
| | RNA 4 | Ribosomal 4 |
| D21S13 | AD1 | Alzheimer Disease 1 (by linkage) |
| D21S95 | | |
| | APP | Amyloid beta (A4) Precursor Protein |
| D21S93 | | |
| | SOD1 | Cu-Zn SuperOxide Dismutase 1, soluble |
| D21S58 | | |
| | IFNAR | Interferon, alpha; receptor |
| | IFNBR | Interferon, beta; receptor |
| | IFNTI | Interferon, gamma; transducer 1 |
| | IFNGR2 | Interferon, gamma; receptor 2 (confers antiviral resistance) |
| | ALS | Amyotrophic Lateral Sclerosis (by linkage to D21S58) |
| | PAIS | Phosphoribosylglycinamide synthetase |
| | PGFT | Phosphoribosylglycinamide formyl transferase |
| | PRGS | Phosphoribosylglycinamide synthetase |
| D21S17 | | |
| D21S55 | | |
| | ERG | Avian erythroblastosis virus E26 oncogene-related |
| D21S57 | | |
| | ETS 2 | Avian erythroblastosis virus E26 oncogene homolog2 |
| | HMG 14 | High mobility group protein 14 |
| D21S3 | | |
| D21S15 | | |
| | MX1, 2 | Myxovirus Influenza resistance, homolog of murine |
| | BCEI | Breast Cancer, estrogen - inducible sequence |
| | CBS | Cystathionine beta synthetase |
| | EPM1 | Progressive Myoclonus Epilepsy (Unverricht-Lundborg type) (by linkage to D21S112) |
| | CRYA1 | Crystallin, alpha polypeptide 1 |
| | PFKL | Phosphofructokinase, liver type |
| | CD18 | Antigen CD18 (p95) Lymphocyte Function |
| | (LFA-1B) | Associated Antigen 1 |
| | COL6A1 | Collagen, type VI, alpha 1 |
| | COL6A2 | Collagen, type VI, alpha 2 |
| | S100B | S100 protein, beta polypeptide (neural); glial (specific protein associated with neurite outgrowth) |

suggests that genes causing mental retardation are present throughout the chromosome, in all regions now defined. However, the size of the duplications now defined is too large to allow the distinction of genes whose duplication may *not* result in MR. Therefore definition of such regions requires the identification of rare individuals with very small duplications. An alternative approach to this dilemma is the study of rare individuals who are deleted for regions of chromosome 21. Since it may be expected that genes with tissue specific deleterious effects when trisomic may also be damaging when monosomic, the study of these deleted individuals may help to define chromosomal regions carrying genes with particularly benign or devastating effects when aneuploid. Our preliminary study of such individuals yielded the unexpected result that deletion of a large 22 Mb region of chromosome 21q11.2-q21 was compatible with normal intelligence (Korenberg et al. 1991c). Recent extension of this study (Korenberg et al.1990b) has demonstrated 4 regions likely to have distinct effects on cognitive abilities when deleted. These include two regions associated with mild defects on normal intelligence and one with moderate psychomotor retardation. Deletion of the fourth region is associated with profound retardation. It may be significant that this region overlaps the region apparently responsible for major phenotypic effects of Down Syndrome. Further molecular analysis of such individuals may provide keys to the genes involved in the mental retardation of Down Syndrome. Although what has been done already is clearly an exciting start, the future goals are equally clear. The size of the regions involved must be reduced through the identification and analysis of further informative cases and the completion of the chromosome 21 physical map. The genes mapping within each region must be identified and their potential roles in development assessed. Finally, but of primary importance, each feature of the phenotype must be defined, at the cellular, physiological, physical and developmental levels. These studies hold the promise of an ultimate understanding of the molecular basis of the components of the Down Syndrome phenotype, including mental retardation, congenital heart disease, immune deficits, risk of leukemia, and the link to Alzheimer disease.

## ACKNOWLEDGMENTS

This work took place in part at SHARE's Child Disability Center and was supported by grants from the Alzheimer's Association and the American Health Assistance Foundation and the Carmen and Louis Warschaw endowment fund. Figures 1 and 4 and Table 1 were modified from Korenberg, 1991a; figures 2 and 3 were modified from Korenberg, 1990a with permission.

## REFERENCES

Burmeister M, Kim W, Roydon-Price E, De Lange T, Tantravahi U, Myers R, Cox D (1990). A Map of the Distal Region of the Long Arm of Human Chromosome 21 Constructed by Radiation Hybrid Mapping and Pulsed-Field Gel Electrophoresis. Genomics 9: 19-30.

Carritt B, Litt M (1989). Report of the Committee on the Genetic Constitution of chromosome 20, 21, and 22. Cytogenet Cell Genet (Human Gene Mapping 10) 51:372-383.

Cox D, Burmeister M, Roydon-Price E, Kim S, Myers R (1990). Radiation Hybrid Mapping: A Somatic Cell Genetic Method for Constructing High-Resolution Maps of Mammalian Chromosomes. Science 250: 245-250.

Daniel A (1979). Normal Phenotype and Partial Trisomy for the G Positive Region of Chromosome 21. J Med Genet 16: 227-235.

Epstein C (1986). Consequences of Chromosome Imbalance: Principles, Mechanisms, and Models. Cambridge Univ Press, New York.

Gardiner K, Horisberger M, Kraus J, Tantravahi U, Korenberg J, Rao V, Reddy S, Patterson D (1989). Analysis of Human Chromosome 21: Correlation of Physical and Cytogenetic Maps: Gene and CpG Island Distribution. EMBO J 9: 25-34.

Korenberg JR (1990a): Molecular Mapping of the Down Syndrome Phenotype. In Epstein C.J. (ed.) "Molecular Genetics of Chromosome 21 and Down Syndrome", New York: Wiley-Liss, pp 105-115.

Korenberg J, Falik-Borenstein T, Muenke M, Mennuti M, Pulst S (1990b). Partial Monosomies of Chromosome 21 and Mental Retardation: Molecular Definition of the Region. Am J Hum Genet 47: A31.

Korenberg J, Kawashima H, Pulst S, Ikeuchi T, Ogasawara N, Yamamoto K, Schonberg S, Kojis T, Allen L, Magenis E, Ikawa H, Taniguchi N, Epstein C (1990c). Molecular Definition of the Region of Chromosome 21 that Causes Features of the Down Syndrome Phenotype. Am J Hum Genet 47: 236-246.

Korenberg JR (1991a) Down Syndrome Phenotypic Mapping. In Epstein, CJ (ed.) "The Morphogenesis of Down Syndrome", New York: Wiley-Liss pp 43-52.

Korenberg JR, Bradley C, Disteche C (1991b). Down Syndrome: Molecular Mapping of the Congenital Heart Disease and Duodenal Stenosis. Am J Hum Genet, in press.

Korenberg JR, Kalousek DK, Anneren G, Pulst SM, Hall JG, Epstein CJ, Cox DR (1991c). Deletion of chromosome 21 and normal intelligence: molecular definiton of the lesion. Hum Genet, in press.

Lehesjoki A, Koskiniemi M, Sistonen P, Miao J, Hastbacka J, Norio R, De La Chapelle A (1991). Localization of a Gene for Progressive Myoclonus Epilepsy to Chromosome 21q22. Proc Natl Acad Sci 88: 3696-3699.

McCormick M, Schinzel A, Petersen M, Stetten G, Driscoll D, Cantu E, Tranebjaerg L, Mikkelsen M, Watkins P, Antonarakis S (1989). Molecular Genetic Approach to the Characterization of the 'Down Syndrome Region' of Chromosome 21. Genomics 5: 325-331.

McClure H, Belden K, Pieper W (1969). Autosomal Trisomy in a Chimpanzee: Resemblance to Down Syndrome. Science 165: 1010-1012.

Niebuhr E (1974). Down's Syndrome: The Possibility of a Pathogenetic Segment on Chromosome 21. Humangenetik 21: 99-101.

Pueschel S, Rynders J (eds) (1982). Down Syndrome. Advances in Biomedicine and the Behavioral Sciences. Ware Press, Cambridge MA.

Rahmani Z, Blouin J, Creau-Goldberg N, Watkins P, Mattei J, Poissonier M, Prieur M, Chettouh Z, Nicole A, Aurias A, Sinet P, Delabar J (1989). Critical Role of the D21S55 Region on Chromosome 21 in the Pathogenesis of Down Syndrome. Proc Natl Acad Sci 86: 5958-5962.

Siddique T, Figlewicz D, Pericak-Vance M, Haines J (1991). Linkage of a Gene Causing Familial Amyotrophic Lateral Sclerosis to Chromosome 21 and Evidence of Genetic-Locus Heterogeneity. N Engl J Med 324:20:1381-1384.

Tanzi R, Haines J, Watkins P, Stewart G, Wallace M, Hallewell R, Wong C, Wexler N, Conneally P, Gusella J (1988). Genetic Linkage Map of Chromosome 21. Genomics 3: 129-136.

Warren AC, Slaugenhaupt SA, Lewis JG, Chakravarti A, Antonarakis SE (1989). A genetic linkage map of human chromosome 21. Genomics 4:579-591.

Zellweger H (1977). Down Syndrome. In: Vinken P, Bruyn G (eds) Handbook of Clinical Neurology, Vol 31, Part II. North Holland Press, pp 367-469.

Zipursky A, Peeters M, Poon A (1987). Megakaryoblastic Leukemia and Down Syndrome - A Review. In McCoy E, Epstein C (eds) Oncology and Immunology of Down Syndrome, pp 33-56.

# Current and Future Modes of Prenatal Diagnosis

Mark H. Bogart

Down syndrome is characterized by an additional chromosome number 21. Prenatal diagnosis of trisomy 21, Down syndrome, is achieved by examination of fetal or placental chromosomes. This can be accomplished by obtaining a small biopsy of the placenta (chorionic villi sampling) or by sampling the amniotic fluid (amniocentesis). Chorionic villi sampling (CVS) is usually performed between 8 to 12 weeks of gestation while amniocentesis is performed from 12 weeks gestation onward. Umbilical cord blood sampling may also be used, but is usually restricted to pregnancies in which there is equivocal data from prior study or a potential fetal abnormality is ascertained during late second or the third trimester. The sampling techniques present some risk to fetal well-being and are, therefore, restricted to pregnancies in which the risk of Down syndrome or other abnormality is sufficiently high to justify the procedure. The risk that is considered "sufficiently high" is arbitrary and has changed over the years. In 1983, the President's Commission for the Study of Ethical Problems in Medicine and Biomedical and Behavioral Sciences recommended that genetic amniocentesis should be available to all pregnant women. However, because the diagnostic procedures are relatively expensive, labor-intensive and limited numbers of trained technicians are available for performing chromosome evaluation, most physicians recommend amniocentesis for those women with a Down syndrome birth risk greater than about 1 in 365. Thus, there is a need for effective screening techniques to determine which pregnancies are at highest risk of having a fetus affected with Down syndrome.

## SCREENING FOR DOWN SYNDROME

Methods for establishing which pregnancies are at highest risk for having a fetus affected with Down syndrome fall into three categories: epidemiology, biochemical parameters and ultrasound findings. Of these, the association between increasing maternal age and increasing incidence of Down syndrome is the best known. This association was first identified by Penrose (1933) and has since been amply confirmed. This association forms the basis of offering prenatal diagnosis to women of advanced maternal age, usually age 35 or greater. The risk of giving birth to a Down syndrome baby is 1 in 384 at age 35 rising to 1 in 112 by age 40 and 1 in 11 by age 48 (Cuckle et al., 1987). However, this advanced age group of women only accounts for about 7% of births each year. Therefore, offering prenatal diagnosis to this 7% of the pregnant population will detect only about 20% to 25% of Down syndrome pregnancies.

Other indications for prenatal diagnostic procedures are: family history of Down syndrome, previous Down syndrome pregnancy or known familial translocation involving chromosome 21. In general, women who have had one Down syndrome pregnancy have a recurrence risk of approximately 1% (Hsu, 1986). However, some women with normal chromosomes, as determined from blood chromosome analysis, have recurrent trisomy 21 conceptions, most likely as a result of germ line mosaicism. Women who are carriers of a balanced translocation involving chromosome number 21 have an increased risk for Down syndrome pregnancies. The risk is approximately 10% to 15% for Robertsonian translocation except the rare case of a familial 21/21 translocation which has a 100% risk for Down syndrome. For a detailed discussion of epidemiologic risk factors for chromosome abnormalities, including Down syndrome, the chapter by Hsu (1986) is recommended.

## BIOCHEMICAL SCREENING PARAMETERS

### Alpha-Fetoprotein

In 1984, Merkatz and coworkers observed an association between reduced levels of alpha-fetoprotein (AFP) in maternal serum and Down syndrome pregnancies. This observation has been confirmed by many subsequent studies. Using a combination of maternal age and AFP levels, detection of about 40% of Down syndrome pregnancies can be achieved when about 6-7% of pregnancies have prenatal diagnosis (Cuckle et al., 1984). On the other hand, at least 60% of Down syndrome pregnancies will not be ascertained using this methodology. Despite this maximum theoretical detection of only 40% of Down syndrome pregnancies, by 1988 approximately 1 million pregnancies annually were provided with Down syndrome risk assessment (Palomaki et al., 1990a) using AFP and age calculations. For detailed discussion of the use of maternal serum AFP in screening for Down syndrome, the review by Knight et al. (1988) is recommended.

### Chorionic Gonadotropin

In 1987, Bogart and coworkers reported a strong association between elevated maternal serum levels of human chorionic gonadotropin (hCG), the free gonadotropin alpha subunit (α-subunit) and pregnancies with a fetal chromosome abnormality, particularly Down syndrome. Human chorionic gonadotropin is composed of two noncovalently bound subunits which are independently produced within trophoblasts. Immunoassay comparison between the levels of hCG, total beta-hCG (free beta plus intact hCG), free α-subunit and free beta-subunit in normal pregnancies and pregnancies with trisomy 21 indicate that there is excess production in some trisomy 21 pregnancies. Studies suggest that pregnancies with elevated levels of hCG or total beta-hCG may also have elevated levels of α-subunit (Bogart et al., 1987; Bogart et al., 1989). These same studies also show that in trisomy 21 pregnancies normal levels of hCG or total beta-hCG may be accompanied by elevated levels of α-subunit or vice versa. Thus, disruption of the normal control of hCG production may occur in the production of either and/or both subunits, and this appears to vary in different pregnancies. It has recently been shown that despite the elevated levels of hCG in trisomy 21 pregnancies, the ratio of immunoactive to bioactive hCG is not different from the ratio in normal outcome pregnancies (Kratzer et al., 1991a).

Most studies have concentrated on the evaluation of hCG or beta-hCG and, overall, indicate that approximately 56% of trisomy 21 pregnancies have levels in excess of 2 multiples of the median (MoM) level of normal pregnancies from the same gestational age (or greater than 95%). These studies include a total of 342 pregnancies with Down syndrome and the weighted average detection is 56% with a weighted average false positive rate of 7.85% (Table 1).

Despite claims of improved detection of Down syndrome (Macri et al., 1990a), free hCG beta-subunit measurement gives essentially the same detection as use of hCG or total beta-hCG (Table 1).

Because the level of hCG in maternal serum changes as gestation progresses, median levels from pregnancies need to be adjusted for each week of gestation prior to week 18 and current practice includes a week-by-week adjustment in the normal median value for weeks 15 to 20 (the time period during which hCG screening occurs). In addition to a gestational age adjustment, there is a correlation between maternal weight and maternal serum levels of hCG showing a decrease in serum hCG concentration with increasing maternal weight (Palomaki et al., 1990b; Bogart et al., 1991). This necessitates a weight adjustment for accurate trisomy 21 risk assessment. Further, there appear to be racial differences

**Table 1.**

| N | >2.0 MoM | > 95% | F.P. | Reference |
|---|---|---|---|---|
| 30 | 63% | | 9.5% | Bogart (1987,89) |
| 77 | 55% | | 10% | Wald (1988) |
| 38 | 50% | | 8.9% | Petrocik (1989) |
| 43 | 44% | | 6.6% | Bartels (1990) |
| 29 | | 58% | 5.0% | Macri (1990a) |
| 60 | | 64% | 5.0% | Muller, Boué (1990) |
| 16 | 68% | | 9.1% | Suchy, Yeager (1990) |
| 49 | 53% | | 8.5% | Crossley (1991) |
| 342 | 56% | | 7.85% | |

in hCG production even after weight adjustment. After weight adjustment, Black patients produce, on average, 9% more hCG than do Caucasian patients and Oriental patients produce 16% more hCG than Caucasians (Bogart et al., 1991). Earlier studies of Black patients that did not include a weight adjustment also found that Blacks have higher serum hCG concentrations than Whites (Muller and Boué, 1990; Simpson et al., 1990; Suchy and Taylor, 1990; Canick et al., 1990a).

Twin pregnancies have higher levels of hCG than do singleton pregnancies (Hussa, 1987; Canick et al., 1990b; Alpert et al., 1990; Bogart et al., 1991). However, at this time, insufficient data exists to confidently establish Down syndrome risk in multiple gestation pregnancies using hCG. Studies suggest hCG concentration may be elevated in patients with diabetes (Barbieri et al., 1986; Canick et al., 1990b). Smoking has been reported to reduce maternal serum hCG levels (Bernstein et al., 1989; Cuckle et al., 1990) and could lead to reduced detection rates in women who smoke.

In pregnancies with fetal trisomy 18 maternal serum hCG levels are usually very low, with most pregnancies having levels less than 0.25 MoM (Bogart et al., 1987; Bogart et al., 1989; Canick et al., 1989; Bartels et al., 1990; Crossley et al., 1990; Suchy and Yeager, 1990; Darnule et al., 1990; Heyl et al., 1990; Ozturk et al., 1990). Similar findings have been reported for amniotic fluid hCG concentrations (Bharathur et al., 1988).

Finally, it is worth noting that pregnancies with abnormal outcomes due to causes other than chromosomal may, on occasion, also have abnormal hCG levels. In particular, 28.8% (13 of 45) of low birth weight babies had hCG levels greater than 2 MoM as did 11.6% of pregnancies with fetal demise, and 15.2% of pregnancies with other types of abnormal outcomes (Bogart et al., 1991). hCG levels are reported to be low in pregnancies affected with anencephaly, though there is no explanation as to why this should be (Canick et al., 1990b).

### Unconjugated Estriol

In 1988, Canick and coworkers reported an association between low levels of unconjugated estriol (uE3) and pregnancies affected with Down syndrome. It was shown that uE3 is a better identifier of Down syndrome pregnancy than AFP and could be used for Down syndrome screening (Wald et al., 1988a). Subsequently, Macri et al. (1990b) have claimed that uE3 levels are not lower in Down syndrome pregnancies than normal pregnancies. When used in combination with AFP and hCG, uE3 is reported to improve Down syndrome detection about 5% (Wald et al., 1988b) though others have had better success (MacDonald et al. 1991). Reduced levels of uE3 have also been reported in pregnancies with fetal anencephaly (Canick et al., 1989; Canick et al., 1990) and with fetal trisomy 18 (Canick et al., 1989).

### Pregnancy Specific Glycoprotein

In 1988, Bartels and Lindemann reported elevated levels of maternal serum pregnancy-specific $\beta_1$-glycoprotein (SP1) in pregnancies with fetal Down syndrome. Subsequent studies have confirmed that SP1 is elevated in Down syndrome pregnancies but SP1 is not as good an indicator of Down syndrome as hCG (Wald et al., 1989; Bartels et al., 1990). The use of SP1 in conjunction with other markers has been reported to be unproductive (Wald et al., 1989; Bartels et al., 1990). On the other hand, Petrocik et al. (1990) suggest SP1 may be a useful additional marker.

## Multiparameter Approaches

Different approaches have been proposed to screen for Down syndrome using various biochemical markers in addition to AFP and are summarized in Table 2. This table indicates the number of Down syndrome pregnancies studied, the detection rate, the false positive rate and the parameters used.

As mentioned earlier, AFP is currently used for Down syndrome risk assessment even though AFP is the least reliable single marker. This is because AFP is already being measured as part of neural tube defect screening programs. All studies so far indicate that hCG (or its subunits) is the best single marker and, thus, will be used in any screening program. Screening protocols have been proposed that use AFP and hCG evaluation either alone (Arab et al., 1988; White et al., 1988, Crossley et al. 1991) or in combination with maternal age (Petrocik et al., 1989; Bogart et al., 1991; Crossley et al., 1991)(Table 2).

The additional incorporation of uE3 (Wald et al., 1988) or α-subunit (Bogart et al. 1987) or SP1 (Petrocik et al., 1990) has also been suggested. The addition of uE3 has received the most attention; however, addition of α-subunit appears the most promising. The methods for establishing "at risk" pregnancies using the biochemical parameters vary from simple cutoffs (Bogart et al., 1987) to

**Table 2.**

| N | Detect | F.P. | Parameters | Reference |
|----|--------|-------|--------------------|------------------|
| 77 | 60% | 6.7% | Age,AFP,hCG | Wald 88 |
| 77 | 60% | 4.7% | Age,AFP,hCG,uE3 | Wald 88 |
| 29 | 62% | 10% | AFP,hCG | Arab 88 |
| 15 | 73% | 4.0% | AFP,hCG | White 89 |
| 26 | 81% | 9.0% | Age,AFP,hCG,uE3 | Osthanondh 89 |
| 38 | 68.3% | 5.0% | Age,AFP,hCG | Petrocik 89 |
| 42 | 57.6% | 7.3% | Age,AFP,hCG,uE3 | Norgaard 90 |
| 16 | 63% | 5.0% | Age,AFP,hCG,uE3 | Heyl 90 |
| 46 | 78.3% | 3.4% | Age,AFP,hCG,SP1 | Petrocik 90 |
| 16 | 62.5% | 4.7% | Age,AFP,hCG | Suchy,Yeager 90 |
| 29 | 72.4% | 5.0% | Age,AFP,β-subunit | Macri 90 |
| 49 | 57% | 5.0% | Age,AFP,hCG | Crossley 91 |
| 54 | 60% | 4.6% | Age,AFP,hCG,uE3 | MacDonald 91 |
| 49 | 65.3% | 7.9% | Age,AFP,hCG | Bogart 91 |
| 29 | 86% | 5.0% | AFP,hCG,α-subunit | Bogart, Jones 91 |

hCG/AFP ratios (Arab et al., 1988; White et al., 1988, Crossley et al., 1991) to various multivariate analysis programs (Wald et al., 1988; Petrocik et al., 1989). There is currently no consensus as to which markers should be used, besides AFP and hCG, or which risk assessment formula should be used.

A new and promising approach for both first and second trimester risk assessment has recently been proposed. This approach generates an "aneuploidy index" using maternal serum hCG, α-subunit and progesterone levels in the first trimester (Kratzer et al., 1991b). A modification of Kratzer's concept for second-trimester screening using AFP, hCG and α-subunit resulted in 86% detection of Down syndrome pregnancies (25 of 29) with a 5% false positive rate (6 of 120 normal outcome pregnancies)(Bogart and Jones, 1991). In essence, Kratzer suggests establishment of Down syndrome risk based on the relative differences of multiple biochemical markers by multiplying two ratios to obtain an "aneuploid index". As with all methods, the detection rate can be increased or decreased depending on the false positive rate desired. For example, if a detection rate of 93% is desired, the false positive rate would be 8.3% whereas 86% detection results in a 5% false positive rate (Bogart and Jones, 1991). It should be remembered that for population screening, the initial false positive rates indicate the percentage of pregnant women referred for ultrasound confirmation of gestational age. For current AFP screening programs, it is reported that about 35% of the women who have ultrasound evaluation have gestational age corrections sufficient to remove them from being considered at risk (Lustig et al., 1988). It is anticipated that similar adjustments will occur in new multiparameter screening programs, thus reducing the initial false positive rate by 1/3. In these circumstances, a 9% initial false positive rate will result in a 6% amniocentesis rate   approximately equal to the number of women over 35. However, instead of detecting 20% of Down syndrome pregnancies as with age alone, the "aneuploid index" method would detect over 90%.

## FIRST TRIMESTER SCREENING

The possibility of using biochemical evaluations for first-trimester screening for fetal Down syndrome is less advanced than second-trimester screening. Milunsky et al. (1988) suggested the possibility of using AFP evaluation because about 30% of first-trimester pregnancies affected with Down syndrome (and other chromosome abnormalities) have AFP levels less than 0.6 MoM. Unconjugated estriol is reported to be lower in Down syndrome pregnancies (Cuckle et al., 1988; Brock et al., 1990) as is SP1 (Brock et al., 1990). hCG levels are reported to be elevated in the first trimester with some studies suggesting that the elevation is not significant (Cuckle et al., 1988; Bogart et al., 1989) while others found greater elevation (Brock et al., 1990; Kratzer et al., 1991b). As mentioned earlier, using hCG, α-subunit and progesterone in the "aneuploid index" approach, 53% (9 of 17) of first-trimester Down syndrome pregnancies were identified while 4.5% (5 of 112) of normal pregnancies were misidentified (Kratzer et al., 1991b).

## ULTRASOUND SCREENING

Because many Down syndrome fetuses have anatomic anomalies, ultrasound evaluation of the fetus can detect some Down syndrome pregnancies. The success of ultrasound detection of Down syndrome varies greatly depending on the equipment used and the skill and experience of the technician or physician. Sonographic signs suggesting Down syndrome are: shortened femur length, shortened humeral length, thickened nuchal fold, duodenal atresia, cardiac defects, cystic hygroma, and hyperechogenic bowel (Benacerraf et al., 1987; Nyberg et al., 1990). Screening for Down syndrome using actual femur length/expected femur length ratio or the BPD/ femur length ratio has been suggested by Benacerraf et al. (1987), but found to be ineffective by others (Lynch et al., 1989; Marquette et al., 1990). Nuchal thickness seems the most promising ultrasound screening parameter with detection rates of 39-75% observed in several small studies (Benacerraf et al., 1987; Lynch et al., 1989; Crane and Gray, 1991). It seems reasonable that pregnancies with any of the ultrasound abnormalities occurring in Down syndrome should be offered amniocentesis. However, routine ultrasound screening of all pregnancies specifically for Down syndrome appears less useful than biochemical screening techniques.

## SUMMARY

Prenatal Down syndrome detection is routinely accomplished via either CVS or amniocentesis. These procedures are offered to the approximately 7% of pregnant women considered to be at highest risk for having a Down syndrome conception. Establishing a woman's risk is possible via a number of biochemical screening approaches, all of which appear most effective in the second trimester. Other factors such as observation of ultrasound abnormalities or a familial balanced translocation present sufficient risk that prenatal diagnosis is recommended regardless of biochemical and/or age-related risk assessment. The currently accepted approach of combined age and AFP risk assessment will soon be expanded to include hCG evaluation. (This expanded approach is currently available at many laboratories but is not yet considered standard of care.) The possibility of incorporation of an additional biochemical parameter in the risk evaluation seems likely, but opinions differ as to whether the additional factor should be $\alpha$-subunit, uE3 or SP1. Most attention has been on the use of uE3 but preliminarily studies suggest that addition of $\alpha$-subunit evaluation gives the best detection rates. These findings suggest detection of about 90% of Down syndrome pregnancies is possible.

## REFERENCES

Arab J, Siegel-Bartelt J, Wong PY, Doran T (1988). Maternal serum beta human chorionic gonadotropin (MSHCG) combined with maternal serum alpha-fetoprotein (MSAFP) appears superior for prenatal screening for Down syndrome (DS) than either test alone. Am J Hum Genet 43: A225.

Barbieri RL, Saltzman DH, Torday JS, Randall RW, Frigoletto FD, Ryan KJ (1986). Elevated concentration of the β-subunit of human chorionic gonadotropin and testosterone in the amniotic fluid of gestations of diabetic mothers. Am J Obstet Gynecol 154: 1039-1043.

Bartels I, Lindemann A (1988). Maternal levels of pregnancy-specific β1-glycoprotein (SP1) are elevated in pregnancies affected by Down's syndrome. Hum Genet 80: 46-48.

Bartels I, Thiele M, Bogart MH (1990). Maternal serum hCG and SP1 in pregnancies with fetal aneuploidy. Am J Med Genet 37: 261-264.

Benacerraf BR, Gelman R, Frigoletto FD (1987). Sonographic identification of second-trimester fetuses with Down's syndrome. N Engl J Med 317: 1371-1376.

Bernstein L, Pike MC, Lobo RA, Depue RH, Ross RK, Henderson BE (1989). Cigarette smoking in pregnancy results in marked decrease in maternal hCG and oestradiol levels. Br J Obstet Gyneacol 96: 92-96.

Bharathur R, Ragam K, Delacruz A, Bircsak M, Haider M, Lee ML (1988). Amniotic fluid beta hCG levels associated with Down syndrome and other chromosome abnormalities. Am J Hum Genet 43#3: A226.

Bogart MH, Pandian MR, Jones OW (1987). Abnormal maternal serum chorionic gonadotropin levels in pregnancies with fetal chromosome abnormalities. Prenat Diagn 7: 623-630.

Bogart MH, Golbus MS, Sorg ND, Jones OW (1989). Human chorionic gonadotropin levels in pregnancies with aneuploid fetuses. Prenat Diagn 9: 379-384.

Bogart MH, Jones OW, Felder RA, Best RG, Bradley L, Butts W Crandall B, MacMahon W, Wians FH Jr, Loeh PV (1991). Prospective evaluation of maternal serum chorionic gonadotropin levels in 3428 pregnancies. Am J Obstet Gynecol (In press).

Bogart MH, Jones OW (1991). Prenatal screening for Down syndrome. Prenat Diagn (In press).

Brock DJH, Barron L, Holloway S, Liston WA, Hillier SG, Seppala M (1990). First-trimester maternal serum biochemical indicators in Down syndrome. Prenat Diag 10: 245-251.

Canick JA, Knight GJ, Palomaki GE, Haddow JE, Cuckle HS, Wald NJ (1988). Low second trimester maternal serum unconjugated oestriol in pregnancies with Down's syndrome. Br J Obstet Gynecol 95: 330-333.

Canick JA, Stevens LD, Abell KB, Panizza DS, Osathanondh R, Knight GJ, Palomaki GE, Haddow JE (1989). Second trimester maternal serum unconjugated estriol and human chorionic gonadotropin in pregnancies affected with fetal trisomy 18, anencephaly, and open spina bifida. Am J Hum Genet 45#4: A255.

Canick JA, Panizza DS, Palomaki GE (1990a). Prenatal screening for Down syndrome using AFP, uE3, and hCG: effect of maternal race, insulin-dependent diabetes and twin pregnancy. Am J Hum Genet 47#3: A270.

Canick JA, Knight GJ, Palomaki GE, Haddow JE (1990b). Second-trimester levels of maternal serum unconjugated oestriol and human chorionic gonadotropin in pregnancies affected by fetal anencephaly and open spina bifida. Prenat Diag 10: 733-737.

Crane JP, Gray DL (1991). Sonographically measured nuchal skinfold thickness as a screening tool for Down syndrome: results of a prospective clinical trial. Obstet Gynecol 77: 533-536.

Crossley JA, Aitken DA, Connor JM (1991). Prenatal screening for chromosome abnormalities using maternal serum chorionic gonadotropin, alpha-fetoprotein, and age. Prenat Diagn 11: 83-101.

Cuckle HS, Wald NJ, Lindenbaum Rh (1984). Maternal serum alpha-fetoprotein measurement: a screening test for Down's syndrome. Lancet 1: 926-929.

Cuckle HS, Wald NJ, Thompson SG (1987). Estimating a woman's risk of having a pregnancy associated with Down's syndrome using her age and serum α-fetoprotein level. Br J Obstet Gynecol 94: 387-402.

Cuckle HS, Wald NJ, Barkai G, Fuhrmann W, Altland K, Knight G, Palomaki G, Haddow JE, Canick J (1988). First-trimester biochemical screening for Down syndrome. Lancet 2: 851-852.

Cuckle HS, Wald NJ, Densem JW, Royston P (1990). The effect of smoking in pregnancy on maternal serum alpha-fetoprotein, unconjugated oestriol, human chorionic gonadotrophin, progesterone and dehydroepiandrosterone sulphate levels. Br J Obstet Gynecol 97: 272-276.

Darnule A, Schmidt D, Weyland B, Greenberg F, Rose E, Alpert E (1990). Serum HCG, AFP and unconjugated estriol levels in trisomy 18 pregnancies in mid-trimester. Am J Hum Genet 47#3: A272.

Heyl PS, Miller W, Canick JA (1990). Maternal serum screening for aneuploid pregnancy by alpha-fetoprotein, hCG and unconjugated estriol. Obstet Gynecol 76: 1025-1031.

Hsu LYF (1986). Prenatal diagnosis of chromosome abnormalities. Genetic Disorders and the Fetus, A. Milunsky ed, Plenum Publishing Corp., 115-167.

Hussa RO (1987). The clinical marker hCG. Praeger Publishers, New York, New York, 97-98.

Knight GJ, Palomaki GE, Haddow JE (1988). Use of maternal serum alpha-fetoprotein measurements to screen for Down's syndrome. Clin Obstet Gynecol 31#2: 306-327.

Kratzer PG, Golbus MS, Finkelstein DE, Taylor RN (1991a). Trisomic pregnancies have normal human chorionic gonadotropin bioactivity. Prenat Diagn 11: 1-6.

Kratzer PG, Golbus MS, Monroe SE, Finkelstein DE, Taylor RN (1991b). First trimester aneuploidy screening using serum human chorionic gonadotropin (hCG), free αhCG, and progesterone. Prenat Diagn (In press).

Lustig L, Clarke S, Cunningham G, Schonberg R, Tompkinson G (1988). California's experience with low MS-AFP results. Am J Med Genet 31: 211-222.

Lynch L, Berkowitz GS, Chitkara U, Wilkins IA, Mehaleck KE Berkowitz RL (1989). Ultrasound detection of Down syndrome: is it really possible? Obstet Gynecol 73: 267-270.

MacDonald ML, Wagner RM, Slotnick RN (1991). Sensitivity and specificity of screening for Down syndrome with alpha-fetoprotein, hCG, unconjugated estriol, and maternal age. Obstet Gynecol 77: 63-68.

Macri JN, Kasturi RV, Krantz DA, Cook EJ, Moore ND, Young JA Romero K, Larsen JW (1990a). Maternal serum Down syndrome screening: free β-protein is a more effective marker than human chorionic gonadotropin. Am J Obstet Gynecol 163: 1248-1253.

Macri JN, Kasturi RV, Krantz DA, Cook EJ, Sunderji SG, Larsen JW (1990b). Maternal serum Down syndrome screening: unconjugated estriol is not useful. Am J Obstet Gynecol 162: 672-673.

Marquette GP, Boucher M, Desrochers M, Dallaire L (1990). Screening for trisomy 21 with ultrasonographic determination of biparietal diameter/femur length ratio. Am J Obstet Gynecol 163: 1604-1605.

Merkatz IR, Nitowsky HM, Macri JN, Johnson WJ (1984). An association between low maternal serum alpha-fetoprotein and fetal chromosome abnormalities. Am J Obstet Gynecol 148: 886-894.

Milunsky A, Wands J, Brambati B, Bonacchi I, Currie K (1988). First-trimester maternal serum α-fetoprotein screening for chromosome defects. Am J Obstet Gynecol 159: 1209-1213.

Muller F, Boué A (1990). A single chorionic gonadotropin assay for maternal serum screening for Down's syndrome. Prenat Diagn 10: 389-398.

Norgaard-Pedersen B, Larsen SO, Arends J, Svenstrup B, Tabor A (1990). Maternal serum markers in screening for Down syndrome. Clin Genet 37: 35-43.

Nyberg DA, Resta RG, Luthy DA, Hickok DE, Mahony BS, Hirsch JH (1990). Prenatal sonographic findings of Down syndrome: review of 94 cases. Obstet Gynecol 76: 370-377.

Osathanondh R, Canick JA, Abell KB, Stevens LD, Palomaki GE, Knight GJ, Haddow JE (1989). Second trimester screening for trisomy 21. Lancet 2: 52.

Ozturk M, Milunsky A, Brambati B, Sachs ES, Miller SL, Wands JR (1990). Abnormal maternal serum levels of human chorionic gonadotropin free subunits in trisomy 18. Am J Med Genet 36: 480-483.

Palomaki GE, Knight GJ, Holman MS, Haddow JE (1990a). Maternal serum α-fetoprotein screening for fetal Down syndrome in the United States: results of a survey. Am J Obstet Gynecol 162: 317-321.

Palomaki GE, Panizza DS, Canick JA (1990b). Screening for Down syndrome using AFP, uE3 and hCG: effect of maternal weight. Am J Hum Genet 47#3: A282.

Penrose LS, (1933). The relative effects of paternal and maternal age in mongolism. J Genet 27: 219-224.

Petrocik E, Wassman ER, Kelly JC (1989). Prenatal screening for Down syndrome with maternal serum human chorionic gonadotropin levels. Am J Obstet Gynecol 161#5: 1168-1173.

Petrocik E, Wassman ER, Lee J, Kelly JC (1990). Second trimester maternal serum pregnancy specific beta-1 glycoprotein (SP1) levels in normal and Down syndrome pregnancies. Am J Med Genet 37: 114-118.

President's Commission for the Study of Ethical Problems in Medicine and Biomedical and Behavioral Science (1983). Screening and Counseling for Genetic Conditions. U.S. Government Printing Office, Washington, D.C.

Simpson JL, Elias S, Morgan CD, Shulman L Umstot E, Andersen RN (1990). Second trimester maternal serum human chorionic gonadotropin and unconjugated oestriol levels in blacks and whites. Lancet 335: 1459-1460.

Suchy SF and Taylor HA (1990). Differences between second trimester maternal serum hCG levels in blacks and whites. Am J Hum Genet 47#3: A285.

Suchy SF, Yeager MT (1990). Down syndrome screening in women under 35 with maternal serum hCG. Obstet Gynecol 76: 20-24.

Wald NJ, Cuckle HS, Densem JW, Nanchahal K, Canick JA, Haddow JE, Knight GJ, Palomaki GE (1988a). Maternal serum unconjugated oestriol as an antenatal screening test for Down's syndrome. Br J Obstet Gynecol 95: 334-341.

Wald NJ, Cuckle HS, Densem JW, Nanchahal K, Royston P, Chard T, Haddow JE, Knight GJ, Palomaki GE, Canick JA (1988b). Maternal serum screening for Down's syndrome in early pregnancy. Br Med J 297: 883-887.

Wald N, Cuckle H, Densem J (1989). Maternal serum specific beta$_1$-glycoprotein in pregnancies associated with Down's syndrome. Lancet 2: 450.

White I, Papiha SS, Magnay D (1989). Improving methods of screening for Down's syndrome. N Eng J Med 320: 401-402.

# Analysis of Non-Disjunction in Human Autosomal Trisomies

Terry Hassold, Sallie Freeman, Carol Phillips,
Stephanie Sherman, and Norma Takaesu

It has been over three decades since Lejeune (1959) first documented the cause-and-effect relationship between an extra chromosome 21 and Down syndrome. Later that same year, Jacobs and Strong described the association between an extra sex chromosome and Klinefelter syndrome (47, XXY). Since those initial reports, we have obtained a great deal of information on the frequency of additional chromosomes (i.e., trisomies) among human conceptions. Approximately 1 in 400 newborns has 47 chromosomes rather than the normal number of 46 chromosomes and among fetuses that miscarry the frequency is much higher—approximately 1 in 4 miscarriages is due to trisomy in the fetus.

In addition to information on incidence, a great deal is known about the impact of trisomy on the human fetus. For example, we know that the severity of the abnormalities depends on the chromosome involved. Thus, the presence of an additional X or Y chromosome (the sex chromosomes) may lead to physical or behavioral problems and, for XXYs, sterility, but typically these individuals are indistinguishable from anyone else in the general population. In contrast, most fetuses with an extra chromosome 13, 18, or 21 miscarry and those that survive to term either die in infancy (trisomies 13 and 18) or have the characteristic features of Down syndrome (trisomy 21). Trisomy for the other chromosomes (1-12, 14-17, 19, 20, 22) has even more severe consequences and inevitably result in fetal death *in utero*.

Despite having this information on frequency and effects of trisomy, we still know very little about the cause of trisomy. That is, why does the fertilized egg occasionally have 47 rather than 46 chromosomes? Part of our ignorance is due to technical and ethical problems associated with studying human eggs and sperm, the two types of cells in which the error leading to trisomy presumably occurs. However another difficulty has been the lack of appropriate molecular tools with which to study chromosome behavior. Recent progress in recombinant DNA methodology has led to identification of highly useful variants, or "polymorphisms", on all human chromosomes, and this now makes it possible to overcome some of these technical problems. In the present report, we summarize our studies of DNA polymorphisms of over 200 fetuses or liveborns with an additional chromosome other than the X or Y chromosome (that is, with an additional "autosomal" chromosome). Our studies demonstrate that errors in the egg are usually the cause of autosomal trisomy but, for some autosomes a significant proportion of trisomies results from errors in the sperm.

## RESEARCH QUESTIONS

We have been applying cytogenetic and molecular techniques to the study of human trisomy, with the ultimate aim to identify underlying mechanisms responsible for non-disjunction. In the present report we summarize our initial studies regarding three important questions:

1) Are there chromosome specific patterns of non-disjunction; e.g., is there variation among chromosomes in the proportion of cases attributable to paternal or maternal meiotic errors?
2) What is the relationship between maternal age and the parent and meiotic stage of origin of trisomy?
3) Is there an association between abnormally high or low levels of genetic recombination and non-disjunction?

For questions 1 and 2, we present our results on trisomies 2, 10, 13, 14, 15, 16, 21, and 22. However, our discussion of recombination (question 3) is restricted to maternal trisomy 21, as we have much more data on this condition than on any other autosomal trisomy.

## METHODOLOGICAL APPROACH

### Study Populations

Trisomic fetuses or liveborns were ascertained from three sources. First, all cases of trisomies 2, 10, 14, 15, 16, and 22 and twelve of the cases of trisomy 13 and eight of the cases of trisomy 21 were ascertained as part of a cytogenetic survey of spontaneous abortions being conducted at Northside Hospital, Atlanta, GA. Over 750 spontaneous abortions have been karyotyped as part of this study (Hassold, unpublished observations), with the frequency and distribution of chromosome abnormalities being similar to those reported in previous studies of spontaneous abortions (Hassold, 1986). Second, 24 trisomy 21 fetuses were ascertained as therapeutic abortions, most of which were studied prenatally because of advanced maternal age. Third, three of the trisomy 13 cases and 107 of the trisomy 21 cases were liveborns, detected as part of newborn screens for chromosome abnormalities or as having clinical features consistent with trisomy.

We were unable to detect any obvious difference in origin of trisomy between fetuses and liveborns for trisomy 13 or 21. Therefore, in presenting the results for these trisomies, we have pooled the results from the different ascertainment categories.

## Cytogenetic Studies

Routine cytogenetic studies of tissue or peripheral blood samples of probands and their parents were done using standard Q- or G-banding procedures. Initially, we also analyzed Q-banded cytogenetic heteromorphisms to determine the parent and meiotic stage of origin of the trisomy. However, in a recent study of Down syndrome individuals, we observed a relatively high rate of error associated with the cytogenetic determinations (Sherman et al., in press). Therefore, in the present study we based our results solely on the DNA marker analyses.

## DNA Studies

DNA samples were extracted from fetal tissue or blood samples and processed for hybridization studies as previously described (Hassold et al., 1985). A total of 54 probes were used to detect RFLPs or VNTRs, including one for trisomy 2 (D2S44), two for trisomy 10 (D10S25, D10S28), eleven for trisomy 13 (RBI, D13Z1, D13S1, D13S2, D13S3, D13S4, D13S5, D13S7, D13S10, D13S39, D13S49), six for trisomy 14 (D14S1, D14S16, D14S19, D14S20, D14S22, D14S23), three for trisomy 15 (D15S24, D15S30, D15S35), eight for trisomy 16 (APRT, CTRB, D16Z2, D16S7, D16S35, D16S36, D16S83, D16S84), sixteen for trisomy 21 (CD18, COL6A1, ETS2, SOD1, D21S1, D21S13E, D21S15, D21S16, D21S17, D21S19, D21S55, D21S58, D21S82, D21S110, D21S12, D21S113) and seven for trisomy 22 (IGLC2, MB, PDGFB, D22S1, D22S9, D22S10, D22S32). Information on the polymorphisms detected by these probes and their physical location is described elsewhere (Kidd et al., 1989). In addition, the D21S13 locus on chromosome 21 was analyzed by PCR in some of the Down syndrome families. The primer sequences and conditions for the PCR reaction and detection of alleles are published (Stinissen et al., 1990).

## Analysis of Recombination in Maternal Trisomy 21

We evaluated the frequency of crossing over between the nondisjoined chromosomes 21 by analyzing all loci for which the mother was heterozygous and determining whether heterozygosity was maintained or reduced to homozygosity in the trisomic offspring. Recombination was considered to have occurred if both reduced and non-reduced loci were identified in the proband (see Table 4 for examples). However, many of the cases with no evidence for recombination had only one or two informative loci. Therefore, we scored trisomies as negative for recombination only when at least three well-spaced markers were informative.

We also assessed heterozygosity/homozygosity to infer the meiotic stage of origin of maternal trisomy 21. For this analysis, we examined the proximal-most marker, and equated cases with a non-reduced proximal marker to meiosis I non-disjunction and those with a reduced proximal marker to meiosis II non-disjunction. We scored only those cases which were informative at one of the following proximally situated markers, D21S16, D21S13E, D21S110, or D21S1; all other cases were considered to be of unknown meiotic origin.

## RESULTS AND DISCUSSION

Are there chromosome-specific patterns of non-disjunction?

Over the past 20 years, cytogenetic heteromorphisms have been used to study the parent and meiotic stage of origin of several thousand trisomic fetuses and liveborns (see Figure 1A for example). These studies indicate that most trisomy is maternally derived, but they also suggest that certain chromosomes may be more susceptible to paternal non-disjunction than others. For example, paternal non-disjunction has been reported to account for 20-25% of cases of trisomy 21 (e.g., Bricarelli et al., 1989) but fewer than 5% of cases of trisomies 3, 4, 14, 15, 16, and 22 (Hassold and Takaesu, 1989).

Recently, two studies have questioned the reliability of the results of the cytogenetic studies of trisomy 21 (Antonarakis et al., 1991; Sherman et al., in press), leading us to use DNA markers to reinvestigate the parental origin of several different autosomal trisomies (see Figure 1B for example). The results of our studies of 256 trisomic fetuses or liveborns, consisting of cases of trisomies 2, 10, 13, 14, 15, 16, 21, and 22 are summarized in Table 1. We were able to determine the parental source of the extra chromosome in 202 of the cases with over 90% (188/202) being maternally derived. Thus, our study is in agreement with the previous cytogenetic studies in attributing most trisomy to maternal non-disjunction.

However, the present study contradicts the previous chromosome heteromorphism studies in two important respects. First, paternal errors accounted for only 5% (6/126) of our cases of trisomy 21, a highly significant difference from the

Table 1. DNA Marker Studies of Parental Origin of Different Autosomal Trisomies (data from Hassold et al., 1988; Hassold and Takaesu, 1989; Sherman et al., in press and unpublished observations)

| Trisomy | Paternal (%) | | Maternal (%) | | Unknown |
|---------|------|--------|-----|---------|---------|
| 2 | 2 | (50%) | 2 | (50%) | 2 |
| 10 | 0 | (0%) | 4 | (100%) | 0 |
| 13 | 2 | (29%) | 5 | (71%) | 8 |
| 14 | 2 | (25%) | 6 | (75%) | 1 |
| 15 | 2 | (25%) | 6 | (75%) | 5 |
| 16 | 0 | (0%) | 36 | (100%) | 12 |
| 21 | 6 | (5%) | 120 | (95%) | 13 |
| 22 | 0 | (0%) | 11 | (100%) | 11 |

A

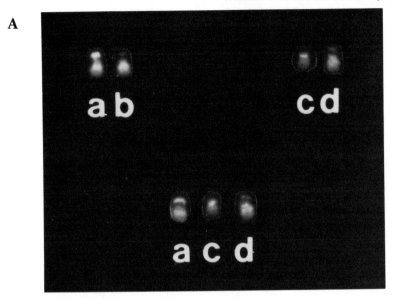

B

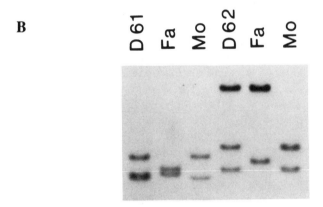

Fig. 1. Approaches to determination of parental origin of trisomy 21: A) analysis of Q-banded chromosome 21 heteromorphisms of father (left), mother (right) and their trisomy 21 offspring (bottom) indicated that the extra chromosome is maternally derived. B) analysis of chromosome 21 DNA polymorphisms (locus D21S112) in two trisomy 21 individulas (D61 and D62) and their parents; in each case, the trisomic individual has three allelels, two of which are maternally derived.

expectation of 20% based on the cytogenetic studies ($X^2_1$ = 18.29; p<0.001). Secondly, in the present series we observed significant variation among trisomies in the likelihood of paternal non-disjunction, but the variation was unlike that observed in the cytogenetic studies. That is, previous studies suggested a high level of paternal non-disjunction for trisomies 13 and 21, but in our series, trisomy 21 had one of the lowest rates of paternal non-disjunction. Indeed, our study suggests that the proportion of paternally derived cases may simply be a function of chromosome size. Thus, 26% (8/31) of trisomies involving large or medium-sized chromosomes (i.e., 2, 10, 13, 14, 15) were paternal in origin, a highly significant increase over the 4% (6/171) value observed for trisomies involving small chromosomes (i.e., 16, 21, and 22) ($X^2_1$ = 20.92; p<0.001).

If this variation is confirmed on a larger series of cases, it suggests a) the existence of mechanisms of paternal non-disjunction restricted to, or more likely to involve, the large chromosomes or b) the existence of mechanisms of maternal non-disjunction restricted to, or more likely to involve, the small chromosomes. As studies of human sperm chromosomes indicate relatively little variation in non-disjunction rates among chromosomes (Martin and Rademaker, 1990), the latter explanation seems more likely to be correct.

## What Is the Relationship Between Maternal Age and the Parent and Meiotic Stage of Origin of Trisomy?

The association between increasing maternal age and trisomy is one of the most important etiological factors in human genetic disease. For trisomy 21, the relationship is well recognized; an exponential increase in risk occurs in the mid-30s, and this risk is still the most compelling reason for offering prenatal diagnosis to older women. The association between maternal age and other trisomies is not as well known, since most such fetuses spontaneously abort. Nevertheless, the age effect extends to these conditions as well. In fact, for women in their 40s the chance of having a conception trisomic for any chromosome is probably at least one in five (Hassold and Jacobs, 1984).

Despite the obvious importance of the maternal age effect, almost nothing is known about its basis. It is usually attributed to abnormalities in pairing and/or disjunction at maternal meiosis I, an intuitively attractive idea since it may take over 40 years to complete this division. However, there is little direct evidence to support this idea and some suggestive evidence against it. Specifically, cytogenetic studies of parental origin of trisomy 21 have consistently shown similar maternal age distributions and means for cases of paternal and maternal origin and for cases of maternal meiosis I and II origin (e.g., Juberg et al., 1983). These observations have led, in part, to the "relaxed selection" hypothesis (Ayme and Lippman-Hand, 1982), which suggests that the age-related increase in the frequency of trisomy may result from an inability of older mothers to reject trisomic offspring. This implies that younger women have an ability to discriminate against chromosomally abnormal conceptuses and that this ability diminishes with increasing age.

We have been interested in using DNA markers to test this hypothesis and, more generally, to characterize the association between maternal age and the parent and meiotic stage of origin of trisomy. Our preliminary results are summarized in Tables 2 and 3.

Table 2 provides mean maternal ages by parent of origin for the 14 paternally and 188 maternally derived cases. As yet, there are too few cases per trisomy to make statistically meaningful comparisons. However, these data show no obvious differences in mean maternal age for paternally and maternally derived cases for any individual trisomy, or for trisomies taken as a group. Therefore, these preliminary results are consistent with the relaxed selection hypothesis.

However, the results of a second analysis are in conflict with one prediction of the relaxed selection model, i.e., that there should be no difference in maternal age for cases of maternal meiosis I or II origin. In fact, for cases of trisomy 21 of maternal origin, inferred meiosis I cases had a significantly elevated mean maternal age by comparison with inferred meiosis II cases (t = 2.48; p< 0.05) (Table 3, see Material and Methods for discussion of meiotic stage determinations in trisomy 21). This suggests that, at least for trisomy 21, the association between

Table 2.  Mean Maternal Age by Parent of Origin of Trisomy

| Trisomy | Parent of Origin (no. of cases) | |
| --- | --- | --- |
| | Paternal | Maternal |
| 2 | $34.5 \pm 0.7$ (2) | $28.5 \pm 6.4$ (2) |
| 10 | — | $36.5 \pm 9.7$ (4) |
| 13 | $34.5 \pm 6.4$ (2) | $33.4 \pm 5.1$ (5) |
| 14 | $33.0 \pm 4.2$ (2) | $30.8 \pm 6.7$ (6) |
| 15 | $33.5 \pm 2.1$ (2) | $37.3 \pm 3.6$ (6) |
| 16 | — | $31.1 \pm 3.9$ (36) |
| 21 | $31.2 \pm 4.5$ (6) | $31.4 \pm 5.8$ (120) |
| 22 | — | $34.8 \pm 5.0$ (9) |

Table 3. Maternal Age Distribution by Presumptive Meiotic Stage of Origin in Trisomies of Maternal Origin

| Presumptive Meiotic Stage of Origin | Total Cases | AGE GROUP | | | | | Age    S.D. |
| --- | --- | --- | --- | --- | --- | --- | --- |
| | | <25 | 25–29 | 30–34 | 35–39 | >39 | |
| Meiosis I | 63 | 6 | 15 | 19 | 19 | 4 | $32.2 + 5.4$ |
| Meiosis II | 20 | 4 | 7 | 5 | 4 | 0 | $29.1 + 4.7$ |

increasing maternal age and trisomy may derive from an age-related increase in meiosis I non-disjunction.

## Is Non-disjunction Associated With Abnormally Low or High Levels of Genetic Recombination?

In yeast and in female Drosophila, the relationship between errors of recombination and non-disjunction is well established. In both, meiotic mutants have been identified which are known to decrease the level of recombination and to increase the level of aneuploidy (e.g., Sandler, 1981; Surosky and Tye, 1988). Additionally, Merriam and Frost (1964) observed that X chromosome non-disjunction in Drosophila is more likely to involve bivalents that have undergone zero or two exchanges than those with one exchange, suggesting that failure to pair or "chromosome entanglement" (Bridges, 1916) due to multiple exchanges can lead to trisomy.

Until recently, there were no data on the possible relationship between recombination and non-disjunction in humans. However, four years ago, Warren et al. (1987) presented direct evidence of a possible association between reduced recombination and trisomy 21. In an analysis of 34 families, they observed virtually no recombination in any chromosome 21 involved in non-disjunction, while recombinants were readily observed among controls. More recently, we have observed a similar effect for 47, XXYs of paternal origin, and have suggested that most such cases derive from meioses in which the X and Y chromosomes fail to pair (Hassold et al., in press).

In this report, we summarize our observations on recombination in maternally derived trisomy 21. Examples of the analytic approach are provided in Table 4 and the data are summarized in Table 5. Of 120 maternally derived

Table 4. Examples of Analysis of Recombination in Six Cases of Maternally Derived Trisomy 21. Loci Known to Be Heterozygous in the Mother Are Scored as "N" (if Heterozygosity Is Maintained in the Trisomic Offspring) or "R" (if Heterozygosity Is Reduced to Homozygosity in the Trisomic Offspring); Dashes Indicate Uninformative Loci. Cases in Which Only "Ns" Are Identified (e.g., Cases 1, 3, 75) Are Consistent With No Crossing-Over Between the Non-Disjoined Chromosomes, Cases With a Single Change From "N" to "R" or "R to N" (e.g., Cases 76 and 77) With at Least One Cross-Over, and Cases With Two Such Changes (e.g., Case 2) With at Least Two Cross-Overs.

| CASE NO. CEN | D21S16 | D21S110 | D21S1 | SOD1 | D21S58 | D21S17 | D21S55 | ETS2 | D21S15 | D21S19 | D21S112 | COL6A1 QTER |
|---|---|---|---|---|---|---|---|---|---|---|---|---|
| 1 | N | - | N | N | - | - | N | - | - | N | N | N |
| 2 | - | - | - | - | N | N | N | - | - | - | R | N |
| 3 | N | - | N | - | N | N | - | N | - | - | N | N |
| 75 | N | - | N | - | - | - | - | - | N | N | N | N |
| 76 | N | - | - | - | R | - | R | - | R | - | R | R |
| 77 | - | R | R | - | - | R | R | - | - | - | N | - |

**Table 5.** Mean Number of Detectable Cross-overs in Presumptive Maternal Meiosis I and II Errors

| Presumptive Meiotic Stage of Origin | Total Cases | No. of Detectable Cross-overs | | | Mean No. of Cross-overs S.D. |
|---|---|---|---|---|---|
| | | 0 | 1 | 2 | |
| Meiosis I | 47 | 32 | 11 | 4 | 0.40 ± .65 |
| Meiosis II | 18 | 4 | 11 | 3 | 0.94 ± .64 |

trisomies, we considered 65 cases to be informative for recombination. These consisted of cases in which we studied at least three well-spaced loci and found no evidence for crossing-over or cases in which one or more cross-overs were identified. We separated the cases into those inferred to be of meiosis I or meiosis II origin and, from Table 5, it is clear that the level of recombination differs between these two categories. The mean number of detectable exchanges for meiosis I cases was less than half that for meiosis II cases, and the overall proportion of cases with/without exchanges was highly significantly different between the two groups ($X^2_1$ = 11.19, p<0.001). Thus, these observations provide strong evidence that abnormalities in pairing and/or recombination at maternal meiosis I are important in the genesis of trisomy 21.

This interpretation is corroborated by the results of a second analysis in which we compared genetic recombination between trisomy 21-generating female meioses and normal female meioses. For this analysis, we constructed a genetic map based on 77 of our cases of maternal trisomy 21 and compared it to a map based on conventional genetic linkage analysis. The methodology used to generate the maps is presented elsewhere (Sherman et al., in press). The two summary maps are shown in Figure 2 and clearly differ in length. The total length of the maternal trisomy 21 map (31 cM) is only one-third that of the normal female map (98 cM), indicating a reduction in genetic recombination in the trisomy-generating meioses.

Thus, in two different types of analyses we observed a significant reduction in recombination in association with maternal trisomy 21. From this we conclude that reduced or absent recombination is a significant contributor to non-disjunction and, indeed, may be one of the most important causes of trisomy 21.

## SUMMARY

We have used DNA polymorphisms to study the origin of the additional chromosome in 256 trisomic fetuses or liveborns. We were able to determine the parental origin in 202 cases, with 188 having an additional maternal and 14 an additional paternal chromosome. In studies of genetic recombination, we observed a significant association between failure of chromosomes to pair and/or exchange material at maternal meiosis I and the genesis of trisomy 21.

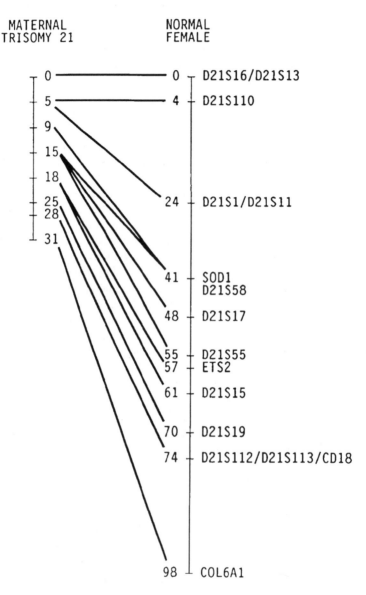

COMPARISON OF GENETIC MAPS (cM)

Fig. 2. Comparison of genetic maps based on female meioses leading to trisomy 21 ('maternal trisomy 21') and female meioses associated with normal segregation of chromosome 21 ('normal female').

## ACKNOWLEDGEMENTS

We wish to thank Jane Hersey for the careful preparation of this manuscript, the staff of the Pathology Department, Northside Hospital, Atlanta, GA for collection of fetal tissue samples, and the many families who participated in this study. This work was supported by NIH grant HD25509 and NIH contract HD92907.

## REFERENCES

Antonarakis SE, Down Syndrome Collaborative Group (1991). Parental origin of the extra chromosome in trisomy 21 as indicated by analysis of DNA polymorphisms. New Engl J of Med 324:872-876.

Ayme S, Lippman-Hand A (1982). Maternal-age effect in aneuploidy: Does altered embryonic selection play a role? Am J Hum Genet 34:558-565.

Bricarelli FD, Pierluigi M, Landucci M. Arslanian A, Coviello DA, Ferro MA, Strigini P (1989). Parental age and the origin of trisomy 21. Hum Genet 82:20-26.

Bridges CB (1916). Nondisjunction as proof of the chromosome theory of heredity. Genetics 1:1-52; 107-163.

Feinberg AP, Vogelstein B (1983). A technique for radiolabeling restriction fragments to high specific activity. Anal Biochem 132:6-13.

Hassold TJ, Sherman SL, Pettay D, Page DC, Jacobs PA (in press). XY chromosome nondisjunction in man is associated with diminished recombination in the pseudoautosomal region. Am J Hum Genet.

Hassold TJ, Takaesu N (1989). Analysis of non-disjunction in human trisomic spontaneous abortions. In: Hassold TJ, Epstein CJ (eds) "Molecular and Cytogenetic Studies of Non-disjunction," New York, Alan R. Liss, Inc., pp. 115-134.

Hassold T, Benham F, Leppert M (1988). Cytogenetic and molecular analysis of sex-chromosome monosomy. Am J Hum Genet 42:534-541.

Hassold T (1986). Chromosome abnormalities in human reproductive wastage. Trends in Genetics 2:105-110.

Hassold T, Kumlin E, Takaesu N, Leppert M (1985). Determination of the parental origin of sex chromosome monosomy using restriction fragment length polymorphisms. Am J Hum Genet 37:965-972.

Hassold T, Jacobs PA (1984). Trisomy in man. Ann Rev Genet 18:69-97.

Jacobs PA, Strong JA (1959). A case of human intersexuality having a possible XXY sex-determining mechanism. Nature, Lond. 183:302.

Juberg RC, Mowrey PN (1983). Origin of nondisjunction in trisomy 21 syndrome. All studies compiled, parental age analysis, and external comparisons. Am J Med Genet 16:111-116.

Kidd KK, Bowcock AM, Schmidtke J, Track RK, Ricciuti F, Hutchings G, Bale A, Pearson P, Willard HF (1989). Report of the DNA committee and catalogs of cloned and mapped genes and DNA polymorphisms. Cytogenet Cell Genet 51:622-947.

Lejeune J, Gautier M, Turpin R (1959). Etude des chromosomes somatiques de neuf enfants mongoliens. C. R. Acad. Sci. Paris 248:1721-1722.

Martin RH, Rademaker A (1990). The frequencey of aneuploidy among individual chromosomes in 6,821 human sperm chromosome complements. Cytogenet Cell Genet 53:103-107.

Merriam JR, Frost JN (1964). Exchange and nondisjunction of the X-chromosomes in female Drosophila melanogaster. Genetics 49:109-122.

Sandler L (1981). The meiotic nondisjunction of homologous chromosomes in Drosophila females. In: de La Cruz F, Gerald P (eds), "Trisomy 21 (Down Syndrome): Research Perspectives," New York, NY, Academic Press, pp. 181-197.

Sherman SL, Takaesu N, Freeman SB, Grantham M, Phillips C, Blackston RD, Jacobs PA, Cockwell AE, Freeman V, Uchida I, Mikkelsen M, Kurnit DM, Buraczynska M, Keats BJB, Hassold TJ (in press). Trisomy 21: Association between reduced recombination and non-disjunction. Am J Hum Genet.

Southern EM (1975). Detection of specific sequences among DNA fragments separated by gel electrophoresis. J Mol Biol 98:503-517.

Stinissen P, Vandenberghe A, Van Broeckhoven C (1990). PCR detection of two RFLPs at the D21S13 locus. Nucl Acids Res 18:3672.

Surosky RT, Tye B-K (1988). Meiotic disjunction of homologs in Saccharomyces cerevisiae is directed by pairing and recombination of the chromosome arms but not by pairing of the centromeres. Genetics 119:273-287.

Warren AC, Chakravarti A, Wong C, Slaugenhaupt SA, Halloran SL, Watkins PC, Metaxotou C, Antonarakis SE (1987). Evidence for reduced recombination on the nondisjoined chromosomes 21 in Down syndrome. Science 237:652-654.

# Learning and Cognition

# Learning and Cognition in Down Syndrome

Lynn Nadel

Sometimes, in thinking about Down syndrome in terms of chromosomes, genes and non-dysfunction, one can forget that there is a person involved, in fact, many people involved. Individuals with Down syndrome are living much longer, more normal lives now, for many of the reasons you will be hearing about in other sessions of this meeting. It is a significant challenge for parents and professionals to help every child with Down syndrome achieve something like his or her real potential. We simply don't know yet what that real potential is, nor how to account for the wide range of differences observed between different children with Down syndrome. The role of scientific research is to inform us more precisely about what is going wrong with brain development, and how this translates into psychological difficulties, in learning, social behavior, or other things.

To put things most simply: brain development is affected by trisomy-21, but much of this happens after birth. The wide variation seen in different children with DS suggests that there is a lot of room for early experiences to influence how the extra chromosome 21 plays out at the behavioral and cognitive level.

The generation of children born with Down syndrome in the past 5-10 years has been psychologically treated rather differently than most who came before. One thing is very clear already: most children with Down syndrome are now doing much better than the average such child used to do. We do not know why this is so, but we need to understand it better. What forms of early experience have produced these benefits?

A first step in making sense of early experience and development in Down syndrome, with the hope of course of translating this understanding into potential improvements in the lives of individuals with DS, is to understand what precise aspects of perceiving, thinking, remembering, talking, understanding, and so on, are altered in DS, and the extent to which these alterations are themselves open to manipulation by particular forms of stimulation.

It is becoming a dogma in cognitive psychology that mental functions are separated into "modules" which reflect the activity of distinct brain systems. Some of the evidence supporting this view comes from the study of various kinds of individuals with brain dysfunctions of one sort or another. The careful study of these people has shown that in most cases certain kinds of functions are disrupted, while others are virtually intact.

The same kind of thing happens in many cases of developmental brain dysfunction, including, we think, Down syndrome. What distinguishes one disorder, say DS, or fragile-X, or Williams syndrome, from another, is the precise brain areas influenced and hence the exact nature of the cognitive and psychological functions that have been affected.

The study of the development of perception, language, and cognition in general, in children with DS has picked up in recent years, offering promise that we will be able to better characterize the range of abilities in children with DS. This will permit moving on to the critical next phase, which is the creation of early stimulation programs that are targeted precisely to the real potential of each child with DS.

At a recent scientific meeting convened by the NDSS a group of researchers reported on their current best understanding of what we know about individuals with DS in terms of cognitive development, language, and other facets of psychological function. Two things seem clear: first, as already noted, the potential of most individuals with DS is far greater than what we had previously thought. This is the good news. Second, there are still real problems and difficulties in particular areas, problems which put serious limits on how parents and professionals can move forward in helping each child with DS.

For example, it is obvious that computer-based remedial programs might be able to help children with DS in a number of ways—programs can be tailored to specific needs of particular children, they can be used repeatedly without fatigue, constantly providing appropriate feedback (something the typical parent or professional simply cannot do indefinitely), they can concentrate on exactly what needs to be practiced.

However, it is premature to get excited about the use of computers as remedial tools with children with DS until we know more about how they can be used most effectively. Some problems are simple—for example, the kinds of movements required for computer use might be difficult for some children with DS, hence it would be necessary to develop appropriate interfaces for these children. Other problems are more difficult—knowing just how fast the program should move, knowing what sorts of motivational problems might arise, and so on, all require more information about the interaction between computers and special populations.

Many aspects of cognitive function in individuals with DS are of interest and importance, but of course language is the area that has received the most attention, and justifiably so. The ability to communicate effectively is essential to social interaction, virtually all forms of gainful employment, and just about every other aspect of modern life. Determining how language in children with DS differs from that in normally-developing children, and what sorts of strategies might help to overcome these problems, is at the top of the agenda for researchers.

Some of the latest information concerning this area is described in the next chapter.

# Development of Speech and Language in Children With Down Syndrome

Jon F. Miller

We have been investigating the development of language and communication skills in children with Down syndrome for the past seven years. The focus of our research has been directed by the question " Are there any unique language learning patterns associated with Down syndrome beyond their general cognitive deficit?" This question suggests a perspective of mental retardation that is contrary to the current unitary view which suggests generally depressed development in all cognitive skills. Recall that mental retardation is a behavioral classification defined by performance on standardized measures of intelligence and adaptive behavior scales. Several hundred different causes of mental retardation have been identified, of which down syndrome is only one. Each of the causes, whether they are genetic, metabolic, trauma or disease process results in unique brain pathology which is the focus of modern neuroscience research. This focus suggests that different brain syndromes should result in different developmental outcomes across the range of mental abilities. The impact of these different brain syndromes on the development of cognitive skills is the focus of the behavioral sciences. Behavioral science research seeks to document and explain the differences in learning and development across children with cognitive deficits resulting from a variety of causes.

Language is the most sophisticated of human cognitive skills. Studies of language learning and use can inform us about the impact of the neurobiological deficits associated with Down syndrome. At the same time detailed behavioral description of language performance of children with Down syndrome can inform molecular genetics where cytogenetic differences exist. A review of Dr. Korenberg's chapter in this volume, suggests that the genes associated with language and communication skills could be mapped if we had a detailed characterization of development in this population. In the next few years we hope to be able to answer the question: " What are the genetic mechanisms responsible for the specific language learning deficits of children with Down syndrome?"

## SPEECH AND LANGUAGE PERFORMANCE

What do we know about the speech and language learning in children with Down syndrome? Reviews of the literature completed recently (Miller, 1987;1988; Fowler, 1990) suggest deficits in both speech and language skill development.

**Speech Production.** The research on speech sound production documents consistent speech intelligibility problems beginning with the appearance of first words and continuing for some children through adulthood. The bases of their persistent unintelligible speech is unclear. At present, no single cause has been identified to explain their speech production patterns. A number of hypotheses have been advanced in the research literature, including speech motor control problems where neuromuscular deficits are thought to impair the rapid coordinated movements of the speech production system, i.e., the lips, tongue, jaw, respiration and phonation. An alternate hypotheses suggests that knowledge of the rules sound combination is delayed or impaired though the research to date only supports greater diversity of sound system knowledge compared to typically developing children. Still another hypothesis suggests that the frequent middle ear infections impair the perception of sound limiting the ability to reproduce accurate speech sounds in production. Given the research to date and our own work on documenting speech motor control deficits, it seems unlikely that a single cause will explain the variety and complexity of speech intelligibility problems exhibited by these children. Work in this area is continuing in several laboratories around the country.

**Language.** Language skills deal with the ability to represent knowledge using words and sentences. Language skills can be expressed orally using speech, or manually through writing, and understood through hearing listening to speech or visually through reading. A child's development language skills must be evaluated through the input side or language comprehension skills as well as the output side or language production skills. The research on the language comprehension skills documents performance consistence with mental age (the non-verbal age equivalent score from a standardized intelligence test) for vocabulary and syntax (the grammatical rules of the language) through the first five years of life. Delays in the comprehension of syntax begin to appear as children are confronted with acquiring complex grammatical features beyond simple sentences. Deficits in acquiring language production skills have been noted consistently for syntax with mixed reports on vocabulary. These deficits appear to get worse with advancing age, suggesting that the rate of language learning in production is much slower that for other cognitive abilities. The over riding question is why language production skills are so much more delayed than language comprehension skills.

## CAUSAL CONSTRUCTS

The causes advanced to account for the language deficits of these children must account for better comprehension than production skills. This leaves out any account concerning hearing or auditory processing deficits, or language input deficits since these skills are necessary for the development of language comprehension. Explanations citing specific cognitive deficits associated with language production skills like failure to access linguistic knowledge, word finding problems or utterance formulation problems hold promise,particularly since a number of studies have found short term memory deficits in this population. In order to explain why children with Down syndrome have deficits in language production, it will be necessary to document how general or specific the deficits, how early they can be detected and to what extent language learning mechanisms available to typically developing children are available to children with DS.

**Summary.** The research data on older children suggests that children with Down syndrome have a unique pattern of language development. Language production is more severely effected than language comprehension with deficits in syntax cited as the primary problem (Fowler, 1990).

## RESEARCH ON EARLY LANGUAGE DEVELOPMENT

Our research is addressing the question of why children with DS have specific language learning problems by investigating the early period of language development. We began by asking how early in the acquisition process the delays in production could be documented. Since vocabulary precedes syntax in development, deficits in vocabulary learning would be evident when first words are expected to appear. Alternatively, the onset of vocabulary acquisition may be on time, but the rate of acquisition may be slower than expected for the child's age or level of cognitive development. Understanding these issues will help us determine if the language learning problems of children with Down syndrome are general, effecting vocabulary, syntax, semantics and the social use of language or specific to syntax as Fowler suggests.

The results of the research published on vocabulary acquisition in children with Down syndrome are somewhat of a paradox. This paradox is primarily the result of experimental and control group matching strategies, some studies matching on mental age (MA) and others on level of general language development. Studies matching children with Down syndrome (DS) and typically developing (TD) children on language production measured by a general index of syntactic development show DS subjects with significantly larger vocabularies than TD children. This could only occur if vocabulary development was more advanced than syntax. The result confirms the asynchrony of syntax and vocabulary acquisition suggesting that the primary deficit is with the development of syntax. Studies matching on MA find children with DS learning fewer words than TD controls in experimental vocabulary learning trials. However, if vocabulary size is compared for language samples of a standard duration, DS and TD produce similar numbers of words.

Clearly, subject matching strategies and the measures used for documenting vocabulary knowledge have contributed to the apparent conflicting results of these studies. In addition, none of these studies matched subject samples on socio-economic status (SES), a factor which is highly correlated with language development. In order to resolve the paradox posed by the current vocabulary studies, we have been investigating early vocabulary acquisition from first words to the onset of syntax, matching on MA and SES, using methods sensitive to total vocabulary size as well as word diversity from a language sample. The remainder of this paper will discuss the results of this work.

## Methods of Investigating Early Vocabulary Development

Measuring early vocabulary knowledge in very young children poses special problems. Children's overall linguistic skill is limited, eliminating direct assessment methods using verbal responses. Two measures have been used in the experimental literature, each measuring something different about the vocabulary acquisition process. The first involves gathering a language sample of a specified length in time, generally a conversation between a parent and child. The sample is transcribed and the number of different words the child produced are counted. This is interpreted as an index of the child's vocabulary diversity and is often generalized as an indication of the child's total vocabulary size. The second method is the number of trials or exposures required by the child to learn a new word under experimental conditions. This measure is interpreted as an index of learning potential and is generalized as an index of the child's rate of learning of new words in the natural environment. Both of these measures have been used to make inferences about the child's overall vocabulary learning.

The best index of vocabulary learning in natural conditions is the child's total vocabulary size. Researchers have been limited to indirect measures as discussed previously, or using diary methods which require special dedication, usually only found in mothers who happen to be linguists. The task of writing down everything the child says through the day for a period of months or years is particularly daunting. Recently an alternative approach has been developed using parent report as an index of child vocabulary knowledge (Bates, Bretherton, & Snyder (1988). This procedure, The Macarthur Communicative Development Inventory, (Fenson, Dale, Reznick, Bates, Thall, Hartung, & Reilly, (1990)) asks parents to check off words they have heard their children produce spontaneously from a list of more than 500 words grouped into major meaning categories. Categories such as animal names, clothing, transportation, etc. This procedure can be completed in approximately 15 to 30 minutes and works quite effectively from the child's production of first words until the child is about two and one half years of age. The validity of parent report as an overall measure of language development, particularly for vocabulary, has been demonstrated in several recent research projects (Dale, Bates, Reznick, & Morisset, 1989; Rescorla, 1989; Beeghly, Jernberg & Burrows, 1989; Tomblin, Shonrock & Hardy, 1989) Bates et al., demonstrated that parent report vocabulary size taken at 20 months of age predicts language achievement at three to 3 and one half years of age, better than direct measures of language achievement at 20 months of age. Further, this measure of parent report has been demonstrated to be quite reliable over time and correlates very highly with vocabulary size as measured in a sample of language taken in conversation with parents. This method allows investigators, for the first time, direct access to the child's total vocabulary.

The Macarthur scale provided us the opportunity to determine if families of children with Down syndrome performed as reliable as families of typically developing children in recording overall vocabulary size. If this proved to be the case, it would allow us the opportunity to compare total vocabulary development of children with Down syndrome with typically developing children at the same mental age and family socioeconomic status. It would also allow us to compare the child's vocabulary production in a language sample with the total vocabulary reported by families to get an idea of how the language sample measure performed as an index of vocabulary size.

## OUR STUDIES OF EARLY VOCABULARY PRODUCTION

We began by comparing the number of different words produced in a language sample for TD and DS children at 11, 14, 17, 20, 23, and 26 months of mental age. Consistent with previous studies by Cardoso-Martins, Mervis, & Mervis (1985) we found no difference between performance by the typically developing group and the group with Down syndrome, except at 20 months of age. These cross-sectional data confirm the vocabulary sizes for subjects matched on mental age and socioeconomic status appear to be no different in the language sample condition.

### Parent Report Measures

Our next task was to determine if families could reliably report their children's vocabulary production. To do this, we compared the number of different words produced in the language sample condition with the number of different words parents reported their child produced spontaneously. Correlations between the language sample vocabularies and the parent report vocabularies for the typically developing children was r = .78 and for the children with Down syndrome was r = .85. Both of these correlations are highly significant and quite robust in accounting for the overall variance of these measures. Our conclusion is that the families of children with Down syndrome perform as well or better than families of typically developing children in accurately reporting their children's vocabulary size. This finding is extremely important for two reasons. First, it documents that parents of children with Down syndrome have a realistic view of their children's developmental progress. Second, the data support the use of the Macarthur scale to monitor vocabulary development in children with Down syndrome.

### Total Vocabulary Size

Having confirmed the reliability of this measure for all of our families, we then went on to compare the total vocabularies of the two groups. As can be seen in Figure 1, the children performed similarly to one another until 17 months of age,

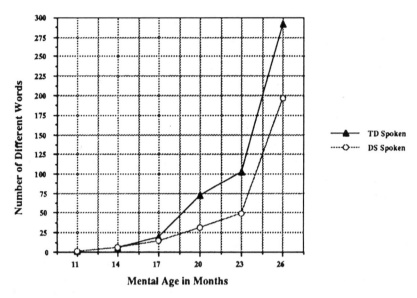

Fig. 1. Number of different words by mental age. Total DS and TD samples.

at which time the typically developing children experienced a rapid acceleration in their acquisition of vocabulary which continues through the 26 months of age. Children with Down syndrome progress at a steady rate until 23 months of age at which time they also experience a period of rapid acceleration. Two things are important to note in Figure 1. First, while there is not a significant difference in total vocabulary size between the two groups, average vocabulary size does begin to diverge for the two groups after 17 months of mental age. Second of all, the children with Down syndrome demonstrate the same period of rapid acceleration of vocabulary observed in typically developing children. This period of rapid acceleration suggests that both groups have the same cognitive mechanisms for vocabulary acquisition though they may be activated at different times and with a somewhat different overall efficiency between the two groups. This research has documented that vocabulary size when measured by language sample method does not yield differences between the two groups overall. But when measures of total vocabulary size are administered, the children with Down syndrome appear to exhibit smaller overall vocabularies after 17 months of mental age though there is considerable variation in vocabulary size in both groups. The question remains, why should children with Down syndrome who are twice the age on average as the typically developing child exhibit smaller vocabularies when presumably they have the same cognitive skills available to them to learn new words. Only longitudinal research on vocabulary acquisition can confirm these preliminary results and begin to address the fundamental question of causal constructs.

## The Acquisition of Sign Vocabulary

Unlike typically developing children, children with Down syndrome are frequently placed in special education classrooms where sign instruction is a part of the curriculum. Sign instruction is introduced by speech language pathologists and special educators as a means of promoting communication development in these children. Sign language is introduced to these children for a variety of reasons, including the belief that signs will be easier for children to learn than words; that gesture is a more natural form of communication for children with Down syndrome than typically developing children, and that children with Down syndrome will exhibit unintelligible speech through the second and third years of life requiring an augmentative communication system to facilitate message transfer. It is assumed that the teaching of sign to these children will enhance their overall communication, that children will spontaneously decrease the use of signs as their oral language becomes more intelligible, that sign use will reduce frustration by making their communication more effective.

There is, surprisingly, very little research on the effect of sign instruction on facilitating communication for children with Down syndrome. Of the many questions to be answered in this area we began by asking about the nature of the sign vocabulary children learned. Do they learn to sign unique words or only those words for which they are already attempting oral production? In order to address this question, we asked parents to indicate those words that their children only produced with a sign, those words their child spoke, those words they produced with sign and speech when completing The Macarthur Communication Development Inventory. This resulted in three lists of words for each child, signed only, spoken only and signed and spoken. The same 44 children with Down syndrome and 46 typically developing children, 11 to 27 months of mental age participated in this study as the previous studies.

The analysis of these data produced some surprising results, (see Figure 2). The sign and oral vocabularies of children with Down syndrome are similar in size at 11 and 14 months of age but at 17 months more than twice as many words are signed as spoken. Sign instruction appears to provide a different vocabulary from their spoken vocabulary which may be an overall "advantage" in vocabulary learning for these children. This sign advantage disappears at 20 and 23 months were spoken in signed vocabularies are the same size again. At 26 months oral vocabulary development accelerates dramatically while sign vocabulary size remains the same or slightly decreases. The number of words that children with Down syndrome both speak and sign is very small in relation to their total vocabularies indicating that they are learning a different set of words in each language.

The overall advantage of sign instruction on vocabulary development can be found by examining the number of different words signed and spoken as a composite measure of total vocabulary and comparing them to the vocabularies of typically developing children. This analysis reveals that there is an increase in

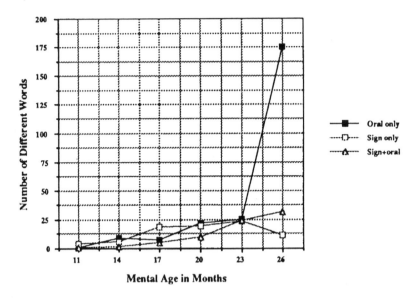

Fig. 2.    Number of different words by mental age. Down sampling using sign.

total vocabulary at all age groups for the children with Down syndrome, (see Figure 3). The vocabularies are more than double the size at 17 and 20 months and increase total vocabulary by a third at 23 months before oral production accelerates, and the sign advantage disappears. The total vocabularies of the children with Down syndrome are actually larger than the typically developing children ages 11 through 17 months, (see Figure 4). After 17 months, however, the typically developing children accelerate dramatically. While sign instruction increases the overall vocabulary size of children with Down syndrome, the average vocabulary is smaller than the typically developing control children at the same mental age and socioeconomic status.

These data suggest that even with two methods of instruction aimed at facilitating vocabulary development in children with Down syndrome, their vocabulary size gets smaller with increasing age when compared to our MA and SES matched typically developing control subjects.

In summary, these data document that vocabulary learning may be a significant problem for children with Down syndrome as they get older. On the one hand, they exhibit vocabularies that are no different from typically developing children of the same mental age and socioeconomic status when measured from a language sample of a standard duration. On the other hand, when examining total vocabulary size from parent report measures, children with Down syndrome exhibit smaller vocabularies than typically developing matched controls as mental age increases. When sign vocabularies are analyzed and added to total vocabulary

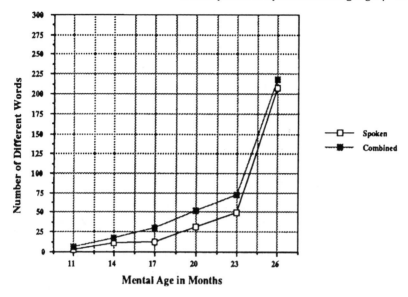

Fig. 3.   Number of different words by mental age. Down sampling using sign.

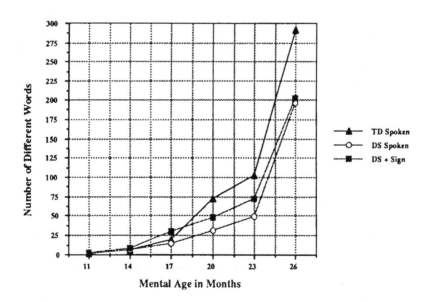

Fig. 4.   Number of different words by mental age. Total DS and TD samples.

size, we find that overall vocabulary size remains smaller for older children with Down syndrome than for the typically developing children.

Two important things emerged from the sign data. One, children are learning unique words in sign, not just sign versions of words they already attempt orally. Second, there is a sign advantage early in vocabulary development giving children a broader variety of opportunities to communicate than they would through speech alone. While overall vocabulary learning remains smaller, clearly early sign instruction, if initiated at the appropriate time, will enhance communication opportunities for children with Down syndrome.

## ACQUISITION STRATEGIES

Do we have any evidence to suggest that children with Down syndrome use similar acquisition strategies as typically developing children? An examination of Figures 1-4 confirms that both groups of subjects exhibit a period of rapid vocabulary learning. That is, vocabulary learning does not progress at the same rate across age in children with Down syndrome and typically developing children. Importance of this observation cannot be overemphasized because it documents that children with Down syndrome have the same mechanisms for understanding that novel words usually represent novel objects in their environment as typically developing children. This ability is known as "fast mapping" in the research literature. They use this fast mapping insight to learn, even through a single presentation of a novel word, that it can in fact stand for a new object, event or relationship. These observations are confirmed by recent research reported by Carolyn Mervis and Jacqueline Bertrand, 1991. Mervis and Bertrand confirm that children with Down syndrome exhibit the same "fast mapping" skills as typically developing children. The question that remains, however, is given similar learning mechanisms, why do children with Down syndrome exhibit smaller overall vocabularies than typically developing children of the same mental age and socioeconomic status? Our longitudinal research will continue to explore this issue.

## PARENT INTRODUCTION AND SUMMARY

Historically, the general development of children with mental retardation was considered a slow motion version of the developmental progress experienced by typically developing children. Mental retardation is a behavioral classification resulting from brain injury due to a variety of causes, genetic, metabolic, trauma and disease processes. Behavioral research is beginning to discover differences in language learning patterns among children with different syndromes. Children with Williams syndrome, for example, exhibit better language an communication skills than would be expected for their general cognitive skills. They exhibit great verbal facility. Children with Down syndrome (DS), on the other hand, have greater difficulty mastering language and communication compared with other cognitive skills. Their language problems, however, are limited to the speaking side of communication. Their listening or understanding of language, develops as expected for their cognitive skills. Our research has examined the nature of the DS child's specific language learning problem.

Over the past six years we have determined that children with DS have strengths in language comprehension, understanding vocabulary and the grammar of the language far better than they are able to produce them in speech. Their speech is complicated by intelligibility problems. Words and sentences are difficult to understand unless you are familiar with the child or what they are trying to say. We have found that if we measure the vocabulary size of children with DS, it is smaller than expected for their mental abilities and their sentences are less grammatically complex than other children of the same mental age. It appears that children with DS have a specific language learning deficit effecting vocabulary and grammar. We have also investigated language learning capacity in these children.

Children with DS are frequently enrolled in programs to teach signs to augment their language production. This is done to overcome their difficulty in producing clear message early in development. We investigated the success of these programs in teaching early vocabulary. We found the children learned a different set of words for signing than for speaking. They learned more total words, sign plus speech, early in development than typically developing children did for speech only. Later in development, after 17 months of mental age, the typically developing children made faster progress than did the children with DS. Even with the advantage of sign production, children with DS had smaller vocabularies than expected for their mental ages.

Both groups of children exhibited a spurt in vocabulary learning where the rate of adding new words increases dramatically. This acceleration in vocabulary learning is an indication that more efficient learning mechanisms are being used by the children to acquire new words. It appears that after children learn about 20 different words, they figure out that words stand for things, actions or relations, and are able to "map" novel words with novel objects. These mechanisms are clearly organized by the children as a product of their learning history. Both the DS and TD children exhibit these abilities, indicating they have similar language learning mechanisms.

What does this research mean for prevention and language intervention programs? Several areas will be reviewed briefly and each should be viewed as a part of an overall program for facilitating communication development.

**Prevention.** There are two concerns to be mentioned here. The first is hearing. About 50% of children with DS have some hearing loss due to ear infections or neurological impairment. Hearing loss has a direct effect on language learning, impairing both understanding and speaking. Routine hearing evaluations should be performed by an Audiologist. Testing should include both the middle ear and inner ear. Children with hearing loss can be helped with amplification through the use of a hearing aid or auditory trainer in school. The second area concerns and changes in the rate of speech and language progress in children with DS. Speech and language is a sensitive index of neurological integrity. Any significant slowing of progress or loss of function should be reviewed by a physician.

**Monitoring.** Our research has demonstrated that the Macarthur Communicative Development Inventory is a valid and reliable tool for monitoring progress in vocabulary development for children with DS through 5-6 years of age. Monitoring developmental progress should be done yearly by a Speech-Language Pathologist familiar with Down syndrome.

**Intervention.** Focus on family based intervention which incorporates frequent opportunities for communication around daily activities, talking about ongoing events and values the child's contribution. Exploit the child's interests in setting the topic for conversation, be responsive to child initiations, responding with your attention, your physical proximity, and verbally repeating and expanding child utterances.

We expect all children with DS to be verbal. Sign instruction early in development can help improve initial communication and reduce frustration. Sign is not a substitute for speech, but a means to augment speech for a short time. Sign use will diminish as speech is more successful in meeting the child's communication needs.

## REFERENCES

Bates, E., Bretherton, I., & Snyder, L. (1988) From First Words to Grammar: Individual Differences and Dissociable Mechanisms. New York: Cambridge University Press.

Beeghly, M., Jernberg, E., & Burrows, E. (1989). Validity of the Early Language Inventory for use with 25-month olds. Paper presented at the Society for Research in Child Development, Kansas City, MO.

Cordoso-Martins, C., Mervis, C. & Mervis, C. (1985). Early vocabulary acquisition in children with Down syndrome. American Journal of Mental Deficiency 90, 177-184.

Dale, P., Bates, E., Reznick, S., & Morrisset, C. (1989) The validity of a parent report instrument of child language at twenty months. Journal of Child Language. 16, 239-250.

Fenson, L., Dale., P., Reznick, S., Bates, E., Thall, D., Hartung, J., & Reilly, J. (1990). Technical Manual for the Macarthur Communicative Development Inventories. Developmental Psychology Laboratory, San Diego State University, San Diego, CA.

Fowler, A. (1990). Language abilities of children with Down syndrome: evidence for a specific syntactic delay. In D. Cicchetti & M. Beeghly (Eds.) Children with Down Syndrome: A Developmental Perspective. Cambridge: Cambridge University Press.

Mervis, C. & Bertrand, J. (1991). Acquisition of new words by children with Down syndrome. Paper presented at the 24th Annual Gatlinburg Conference on Research and Theory in Mental Retardation and Developmental Disabilities, Key Biscayne, Florida.

Rescorla, L. (1989) The language development survey: A screening tool for delayed language in toddlers. Journal of Speech and Hearing Disorders. 29, 394-399.

Tomblin, B., Shonrock, C., & Hardy, J. (1989). The concurrent validity of the Minnesota Child Development Inventory as a measure of young children's language development. Journal of Speech and Hearing Disorders, 54, 101-105.

# Clinical Advances

# The Person With Down Syndrome: Medical Concerns and Educational Strategies

Siegfried M. Pueschel

Many previous reports in the medical literature indicate that severe developmental delay and profound mental retardation are the hallmark of phenotypic expression in persons with Down syndrome. During the past decades, however, the once dreary outlook which implied that individuals with Down syndrome are incapable of being contributing members of society has changed markedly.

Recent advances in both the medical and behavioral sciences have brought about new and more effective approaches in the care and education of handicapped children. For example, comprehensive health maintenance, proper treatment of infections and congenital heart disease, as well as the impact of early intervention and special education services, all have improved the quality of life of children with Down syndrome considerably so that today they can accomplish much more than whatever was previously anticipated.

As contemporary society has attempted to adjust to these evolutionary processes, enlightenment about Down syndrome has resulted in positive attitudinal changes and in a greater acceptance of children with this chromosome disorder.

## INITIAL COUNSELING

All parents who have been told that their child has Down syndrome describe a sensation of overwhelming disbelief. One parent mentioned "it feels as though the world is coming to an end." It is hard for parents to listen further to words and explanations which the physician offers because all their energies are absorbed in the feelings that people have about mental retardation and Down syndrome. Some people try to escape the reality by hoping that some mistake has been made, that the chromosome test would prove the doctor wrong and that their child will be an exception. Even though parents may be told that Down syndrome occurs once in every 800 to 1200 births, they cannot understand why it happened to them. Most people search for an explanation within their own personal behavior. They search

53

for something that happened or something that may have been overlooked. They may think having a child with mental retardation may reflect on their own competence in some way and that other people might think less of them if they have given birth to a child with mental retardation.

Thus, after the birth of a child with Down syndrome, it is of utmost importance that parents be approached by professionals with tact, compassion, and truthfulness. The tone of the atmosphere that will prevail in future years is set during the initial counseling sessions and the way parents are counseled during this initial time period will have a vital influence on their subsequent adjustment. If parents are approached in a supportive and positive manner and if they are told that the infant, although having Down syndrome, is first and foremost a human being, then they will also see positive attributes in their infant and their profound distress will be lessened. During this initial period of adjustment, the parents' general human and parental competence should be stressed; the critical role of parenting needs emphasis as does the need to provide the infant with a chance to be nurtured and loved. If the physician introduces the infant with Down syndrome as a human being, with less emphasis on the features of the syndrome, this will endow the infant with significance and worth (Pueschel & Murphy, 1976).

Although professionals should explain to both parents the meaning of Down syndrome, the anticipated development of the child, the chromosome aberration and related issues as well as provide guidance and support, other parents having an older child with Down syndrome can also be helpful to the new parents. Such resource parents who have been through similar emotional stress when their child with Down syndrome was born, are often more sensitive and can assist the new parents to adjust and to cope more effectively. In a way, the resource parents are living proof that one can survive such a crisis. They can communicate to the new parents that there can be true happiness in a family with a child with Down syndrome.

For most people, it may take weeks or months before they regain a sense of their usual self and before they can pursue their normal routines. The feelings of sadness and loss may never completely go away. Many people describe some beneficial effects of such an experience. They feel that they gain a new perspective on the meaning of life and a sensitivity to what is truly important in life. Frequently, an initial shattering experience such as the birth of a child with Down syndrome can serve to strengthen and unify a family.

## MEDICAL CONCERNS

Within the framework of this paper, it is not possible to cover all clinical disorders observed in persons with Down syndrome in an encyclopedic fashion. Only the most important medical concerns can be discussed briefly. Previous presentations of this topic have detailed various clinical conditions (Pueschel 1987, 1990) and subsequent papers in this volume will focus on specific biomedical aspects in Down syndrome. Therefore, I shall primarily focus on the basic facts and the implications of some frequently observed medical conditions. Moreover, I will emphasize that if we pay attention to these disorders and if we provide appropriate treatment, persons with Down syndrome can achieve a better quality of life.

During the neonatal period, certain congenital anomalies of infants with Down syndrome require immediate attention. Some of them may be life threatening and need to be corrected at once. Others may become apparent during subsequent days and weeks.

## Congenital Cataracts

Congenital cataracts are know to occur more frequently in children with Down syndrome than in the general population. Approximately 3% of newborn children with Down syndrome have dense cataracts which must be extracted soon after birth in order to allow light to reach the retina and to prevent blindness. Subsequently, appropriate correction with glasses or contact lenses will assure adequate vision.

## Congenital Heart Disease

After the diagnosis of Down syndrome has been made in the newborn nursery, infants with this chromosome disorder should be examined by a pediatric cardiologist and an echocardiogram should be obtained. Between 40% and 50% of children have congenital heart disease, the most often observed cardiac lesion is an atrioventricular canal. Other congenital heart defects include ventricular septal defect, tetralogy of Fallot, patent ductus arteriosus, and atrial septal defect. It is of importance to diagnose congenital heart disease during early infancy because some children with severe defects may develop pulmonary artery hypertension, heart failure, and may thrive poorly. If appropriate medical management and prompt surgical repair of congenital cardiac defects are carried out at the optimal time of the child's life, then the quality of life of the individual with Down syndrome will be improved significantly.

## Congenital Anomalies of the Gastrointestinal Tract

It has been estimated that up to 12% of children with Down syndrome have anomalies of the gastrointestinal tract including tracheoesophageal fistula, pyloric stenosis, duodenal atresia, annular pancreas, aganglionic megacolon, and imperforated anus. Most of these anomalies require immediate surgical intervention to allow nutrients and fluids to be absorbed. No form of treatment should be withheld from any child with Down syndrome that would be given unhesitatingly to a child without this chromosome disorder.

## Seizure Disorders

Seizure disorders are known to occur at a greater frequency in children with Down syndrome. We studied the prevalence, onset, and type of seizure disorders, as well as seizure control in a large cohort of 405 individuals with Down syndrome (Pueschel, Louis & McKnight, 1991). We found that 33 (8.1%) of 405 persons with Down syndrome had seizure disorder. With regard to the onset of seizures, a bimodal distribution was noted: 40% of patients began having seizures before the age of 1 year, and another 40% started with seizure activity in the third decade of life. In the younger age group, primarily infantile spasms and tonic-clonic seizures with myoclonus were observed, and the older patients often had partial simplex or partial complex seizures as well as tonic-clonic seizures. It is important to recognize seizure disorders in persons with Down syndrome and if identified, prompt treatment with anticonvulsive medications should be forthcoming.

## Visual Impairment

Children with Down syndrome often have eye disorders including blepharitis, strabismus, keratoconus, nystagmus and refractive errors. Because children with Down syndrome have a high prevalence of ocular problems, they should be examined regularly by a competent pediatric ophthalmologist. Normal visual acuity is important for any child. However, if a child is mentally retarded, an additional handicap of sensory impairment will further limit the child's overall functioning and may prevent the child from participating in significant learning processes.

## Hearing Deficits

It has been reported that up to 80% of children with Down syndrome have some form of hearing impairment (Balkany et al., 1979). Structural abnormalities in the hypopharynx, Eustachian tube dysfunction, abnormalities of the tensor veli palatini muscle, middle ear infections, and fluid accumulation in the middle ear all may result in hearing impairment. Since a hearing deficit in young children with Down syndrome may affect their psychological development, proper assessment of the child's hearing and prompt treatment if a hearing loss is uncovered are of paramount importance. It has been well documented that even a mild conductive hearing deficit may lead to a reduced rate of language development and secondary interpersonal problems.

## Thyroid Dysfunction

We have found that 15% to 20% of children with Down syndrome have some form of thyroid dysfunction (Pueschel & Pezzullo, 1985; Pueschel et al., 1991). If thyroid dysfunction is not recognized early, it may further compromise a child's central nervous system functioning. Because clinical symptoms of hypothyroidism are sometimes interpreted as being part of the "Down syndrome Gestalt," thyroid function studies including T4, T3, and TSH are indicated and should be carried out at regular intervals. If a person with Down syndrome is found to have hypothyroidism, thyroid hormone treatment is indicated. In order that normal learning processes can take place, optimal thyroid function is paramount.

## Skeletal Anomalies

Patellar subluxation, hip dislocation and dysplasia, foot problems, and atlantoaxial instability, all are known to occur at a higher frequency in persons with Down syndrome. Most of these conditions will be discussed in a subsequent chapter. I should like to emphasize, however, that atlantoaxial instability in individuals with Down syndrome should be identified as early as possible because of its relatively high prevalence and its potential for remediation. Because many individuals with Down syndrome may have difficulty verbalizing specific complaints relating to neck discomfort and neuromotor dysfunction, a neurologic examination and roentgenographic studies should be forthcoming. Sometimes, motor disabilities and gain difficulties observed in persons with Down syndrome may conceal significant neurologic concerns. Therefore, it is important to diagnose children with atlantoaxial instability early since a delay in recognizing this condition may potentially result in irreversible spinal cord damage.

## Nutritional Concerns

Some infants with Down syndrome, in particular those with congenital heart defects, thrive poorly and their weight gain is slow. A high caloric intake and a balanced diet are important for these infants.

Later on, however, increased weight gain is often observed in many youngsters with Down syndrome. Therefore, it is important to inform parents with regard to appropriate dietary intake from early childhood on in order to avoid excessive weight gain. A proper intake of a balanced diet, avoidance of high-caloric food items, and regular physical exercises are important for all children including those with Down syndrome.

## Sleep Apnea

Some persons with Down syndrome snore during sleep, have episodes when they do not breathe, are restless sleepers and then may be tired during daytime. These children may suffer from sleep apnea and may have secondary cerebral hypoxia. They may develop pulmonary artery hypertension with resulting cor pulmonale and heart failure. Many persons with Down syndrome have a narrow upper airway which is often due to smaller mid facial bones, increased lymphoid tissue and submucous fat tissue in obese persons. Children with significant upper airway obstruction can often be treated successfully by tonsillectomy and or adenoidectomy, a modified pharyngopalatal surgical approach (Strom, 1986) or other procedures.

## BEHAVIORAL AND PSYCHIATRIC DISORDERS

In recent years, we have observed many children with Down syndrome who displayed conduct disorders, adjustment reactions or depressive episodes. This led us to study the behavioral characteristics of a large group of children with Down syndrome using the Achenbach Child Behavior Checklist. Numerous parents who completed the checklist indicated that their child with Down syndrome often had difficulties concentrating, was impulsive, had trouble sleeping, demanded a lot of attention, exhibited hyperactivity, was stubborn at times and disobedient at school and home (Pueschel 1991).

In another study, we investigated the prevalence and nature of psychiatric disorders in 497 individuals with Down syndrome. The overall frequency of psychiatric disorders in our study population was 22.1%. Youngsters below 20 years old often displayed disruptive behaviors, anxiety disorders, and repetitive behaviors, whereas persons with Down syndrome 20 years and older in addition exhibited depressive disorders (Myers & Pueschel, 1991).

We emphasize to parents the importance of appropriate child rearing practices and proper discipline from early childhood on. If behavioral or psychiatric disorders are identified in persons with Down syndrome, counseling and treatment should be forthcoming.

## EDUCATIONAL STRATEGIES

Appropriate educational strategies should be pursued throughout the child's life starting in early intervention programs and later during preschool, elementary, and high school education as well as during vocational training.

Early intervention tends to enhance the infant's sensory, motor and social developments. There is agreement that it is the quality rather than the sum of total stimulation that shapes the physical and mental developments of the young child with Down syndrome. In addition, preschool plays an important role in the life of the young child by providing social interaction and development and selfhelp skills.

When children with Down syndrome enter school, we often wonder what they will get out of their educational experience. We hope that school will provide the kind of stimulating and rich experiences in which the world appears as an interesting place to explore. Learning situations at school should help the child with Down syndrome gain a feeling of personal identity, self-respect, and enjoyment. School should give a foundation for life and encourage the development of basic academic skills, physical abilities, selfhelp skills, and social as well as language competence. If the school approaches the education in terms of humanizing the teaching process, treating each student as a person with individual integrity, and exposing the child to forces that will contribute to self-fulfillment in the broader sense, then the child with Down syndrome will be given the opportunity to develop optimally in the educational setting.

Young persons with Down syndrome also need to be prepared to function well in the world of work. A vocational program should be designed to provide them with a structured and realistic vocational experience that will enable the student to achieve maximum success in future employment. Such a program should be based on the vocational needs and interests of the student and should emphasize the development of interpersonal skills required in most jobs. Persons with Down syndrome often impress their employers with skills and attitudes that they are not able to verbalize when seeking employment. Proper employment placement will provide a person with Down syndrome with a feeling of self-worth and of making a contribution to society.

## CHARACTERISTICS OF THE PERSON WITH DOWN SYNDROME

I do not want to stereotype the person with Down syndrome, but there are common assets observed in many youngsters which include a natural spontaneity, a genuine warmth, a penetrating clarity of relating to other people, a gentleness, patience and tolerance, to be completely honest, and to engage in unfettered enjoyment of life's gifts (Wolfensberger, 1988). Wolffe (writer for the ABC show LIFE GOES ON) once said "Persons with Down syndrome in some way reflect our own humanity back at us and only our limitations cause us to fail to receive the gifts. There is a goodness, humanity, and magic in these persons that must be protected and never be betrayed." The value of a person with Down syndrome is intrinsically rooted in his very humanity, in his uniqueness as a human being. Yet, the lives of individuals with Down syndrome will remain vulnerable and tentative unless their value is acknowledged as being intrinsic to their humanity.

I do believe in the tremendous potential within each person with Down syndrome, a potential that can be reached if we are sincere in providing the individual with optimal medical, educational, and vocational services and if we are dedicated to enhancing the fullness of life for every person with Down syndrome.

## REFERENCES

Balkany TJ, Downs MP, Jafek BW, Krajicek MJ (1978). Otologic manifestations of Down's syndrome. Surgical Forum 29:582-585.

Myers BA, Pueschel SM (1991). Psychiatric disorders in persons with Down syndrome. The Journal of Nervous and Mental Disease (in press.)

Pueschel SM (1987). Health concerns in persons with Down syndrome. In SM Pueschel, C Tingey, JE Rynders, AC Crocker, DM Crutcher (Eds) New perspectives on Down syndrome. Paul H. Brookes Publishing Co., Baltimore, MD.

Pueschel SM (1990). Clinical aspects of Down syndrome from infancy to adulthood. American Journal of Medical Genetics Supplement 7:52-56.

Pueschel SM, Bernier JC, Pezzullo JC (1981). Behavioral observations in children with Down syndrome. Journal of Mental Deficiency Research (in press).

Pueschel SM, Jackson IM, Giesswein P, Dean MK, Pezzullo JC (1991). Thyroid function in Down syndrome. Research in Developmental Disabilities (in press).

Pueschel SM, Louis S. McKnight P (1991). Seizure disorders in Down syndrome. Archives of Neurology 48:318-320.

Pueschel SM, Murphy A (1976). Assessment of counseling practices at the birth of a child with Down's syndrome. American Journal of Mental Deficiency 81:325-330.

Pueschel SM, Pezzullo JC (1985). Thyroid dysfunction in Down syndrome. American Journal of Diseases of Children 139:636-639.

Strome M (1986). Obstructive sleep apnea in Down syndrome children: a surgical approach. Laryngoscope 96:1340-1342.

Wolfensberger W (1988). Common assets of mentally retarded people that are commonly not acknowledged. Mental Retardation 26:63-70.

# Cardiorespiratory Problems in Children With Down Syndrome

Langford Kidd, MD

Until recently, cardiorespiratory problems and infection were the major causes of premature death and disability in children born with Down Syndrome. This situation has changed with the introduction of appropriate antibiotics, the recognition of the interplay between the lungs and the pulmonary circulation, and the development of surgical techniques for the repair of complex heart conditions during infancy.

Cardiovascular disease in Down Syndrome represents a significant part of the spectrum of congenital heart disease seen by a pediatric cardiologist. In a case-controlled population-based study of infants presenting with congenital heart disease in the first year of life, the Baltimore Washington Infant Study, chromosomal abnormalities accounted for 12% of the total, and other inherited syndromes 8%. The most frequent contributor to the chromosomal abnormalities group was Down Syndrome, representing in fact, 10% of all cases presenting in the first year of life.

When the frequency of various types of congenital heart disease are considered in the general population, ventricular septal defect, as an isolated lesion, is by far the most frequent, representing about 26-28% of the total in both the BWIS study, (Ferencz et al., 1987) and in a clinic-hospital experience. (Keith, 1978) (Table 1.) Atrioventricular canal defect is much rarer, and represents but 3-8% of the total. However, in children with Down Syndrome, 46-62% of whom have congenital heart disease, (Greenwood and Nadas 1976) the frequency of types of defect found is different. In the BWIS study atrioventricular canal defect represented 56% of the total, and ventricular septal defect 19%. (Ferencz et al., 1987), while Park et al., 1977 reported 43% atrioventricular canals and 32% ventricular septal defects.

Left heart obstructive diseases (aortic stenosis, hypoplastic left heart syndrome, coarctation,) and transposition of the great arteries are very rare. Marino et al., (1990a) pointed out that the atrioventricular canal defect in Down Syndrome is more often of the complete variety than in children who do not have Down

Table 1. Frequency of Types of Congenital Heart Disease in a Completely Ascertained Birth Cohort of Infants Diagnosed in the First Year of Life (BWIS) and From a Cardiac Clinic/Hospital Series. (Keith) and From BWIS Patients With Down Syndrome.

| | FREQUENCY PERCENT | | |
| --- | --- | --- | --- |
| | BWIS | KEITH | DOWN SYNDROME |
| VSD | 28 | 28 | 19 |
| AVC | 8 | 3 | 56 |
| ASD | 8 | 6 | 9 |
| PS | 8 | 10 | 0 |
| TOF | 7 | 10 | 6 |
| TGA | 5 | 5 | 0 |
| COA | 5 | 5 | 0 |
| AS | 5 | 7 | 0 |
| PDA | 3 | 10 | 4 |

Syndrome, and that associated malformations are much more common in association with atrioventricular canal in the non-Down group. They further pointed out (Marino et al., 1990b) that the isolated ventricular septal defect in children with Down Syndrome is most often of the inlet type while muscular and subpulmonary ventricular septal defects are rare.

Atrioventricular canal defect is classified by Clark and Takao (1990) as representing a defect in the migration of the extracellular matrix, while Wright et al., (1984) reported abnormal cell adhesiveness in Down Syndrome.

## CIRCULATORY DISTURBANCES IN DOWN SYNDROME IN CHILDREN

The vast majority of the cardiac defects atrioventricular canal defect, ventricular septal defect, atrial septal defect, and persistent ductus arteriosus in children with Down Syndrome bring about left-to-right intracardiac shunting. The most common atrioventricular canal defect involves the endocardial cushions which fail to fuse in the center of the heart. The result of this is that in complete atrioventricular canal defect, there is a large posterior ventricular septal defect, a large septum primum atrial septal defect in the lower part of the atrial septum and a common atrioventricular valve. As pulmonary vascular resistance falls after birth, left-to-right shunting from the left ventricle into the right ventricle, and often across the incompetent mitral and tricuspid valves into the right atrium occurs. There is also mitral regurgitation from the left ventricle into the left atrium and left to right shunting across the primum atrial septal defect. A result of the increased pulmonary blood flow is equal pressures in the right and left ventricles with pulmonary hypertension.

This large left-to-right shunting can lead to congestive heart failure as the pulmonary vascular resistance falls in the 6th to 8th week of life. With the passage of time, continued pressure and flow into the abnormal pulmonary circulation will result in the so-called Eisenmenger syndrome with progressive pulmonary vascular obstructive disease. This will eventually lead to shunt reversal and the appearance of cyanosis.

## CLINICAL PRESENTATION

It was formerly the case that the diagnosis of heart disease in Down Syndrome was not routinely made in early life. However, with the advent of echocardiography and increased awareness, this is no longer the case. A recent study by Schneider et al. (1989) demonstrates that in the first 4 months of life, heart disease in patients with Down Syndrome is recognized at a faster rate that children without Down Syndrome. Indeed, with the use of fetal echocardiography, Down Syndrome can be diagnosed antenatally.

Children with heart disease and Down Syndrome rarely have any problems in the newborn period. This is because the pulmonary vascular resistance is high, and while the pulmonary arterial pressure would also be high, the volume of the left to right shunt is small. Clinical examination at this time may fail to point to the diagnosis of heart disease. This is because, despite the major cardiovascular malformation, there is often no murmur, and unless the auscultator has considerable skill and experience, the subtle changes in the second heart sound may not be obvious.

At 4-6 weeks after birth, the child may present with rapid heart rate and rapid breathing; at this time on physical examination, the liver may be enlarged and the lungs congested, and the increased pulmonary blood flow is causing congestive heart failure. Failure to gain weight frequently accompanies these symptoms. If, however, the fall in pulmonary vascular resistance is slight, the children may progress straight to the Eisenmenger syndrome and be growing and developing normally, but with increased pulmonary vascular resistance the diagnosis may be unsuspected until the appearance of cyanosis, first of all with crying and exercise and then rest, in early childhood.

On physical examination, as remarked above, there may be no murmur, (except of course in tetralogy of Fallot). The second heart sound, however, is loud and single. The electrocardiogram will be abnormal. An atrioventricular canal defect

Table 2. Cardiac Pathophysiology in Down Syndrome

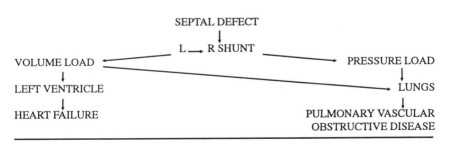

will show left axis deviation and right ventricular hypertrophy; with a ventricular septal defect in the perimembranous area, left ventricular hypertrophy; with an atrial septal defect, in the secundum position, right ventricular hypertrophy and with a ductus, left ventricular hypertrophy. Echocardiography has revolutionized noninvasive diagnosis of this condition.

## PROGRESSIVE PULMONARY VASCULAR OBSTRUCTIVE DISEASE

Progressive pulmonary vascular obstructive disease has been recognized since the time of Eisenmenger (1897) and the usual course of development of pulmonary vascular obstructive disease in patients with large left to right shunts and pulmonary hypertension is that the disease in the pulmonary arterioles is usually well established by the age of 2 years, and that closure of the defect and removal of the stimulus due to the disease does not change the progressive course of the development of obliterative changes in the pulmonary vasculature. (Friedli et al., 1974, DuShane et al., 1976).

The clinical impression that the Eisenmenger reaction was more prevalent and occurred earlier in Down Syndrome was first raised by Chi and Krovitz in 1975, who reported that equal sized left to right shunt in Down and non-Down Syndrome children were associated with higher pulmonary artery pressures in those with Down Syndrome, and these data were supported by Greenwood and Nadas (1976). Yamaki et al. in 1983 reported early intimal and medial changes seen in the pulmonary arterioles of patients with congenital heart disease and Down Syndrome, and in 1984, they similarly reported alveolar hypoplasia, and suggested that this was a potential cause of respiratory failure in Down Syndrome in the early postoperative period. Clapp et al. in 1988 in elegant physiologic experiments reported a significant incidence of early pulmonary vascular obstructive disease in Down Syndrome.

The pathophysiologic basis for this abnormality was established by Cooney and Thurlbeck in 1982 when they reported that examination of the lungs of patients with Down Syndrome demonstrated pulmonary hypoplasia. There was inadequate alveolarization of the terminal lung units distal to the respiratory bronchioles and an absolute reduction in the total number of alveoli. There is also a reduction in the cross-sectional area of the pulmonary vascular bed. The pulmonary capillary network appeared immature, with retention of a double capillary network and fetal structure. It therefore appears that the child with Down Syndrome is predisposed to the development of pulmonary vascular obstructive disease.

## ROLE OF ALVEOLAR HYPOVENTILATION AND OBSTRUCTIVE SLEEP APNEA

That alveolar hypoventilation may occur in Down Syndrome has been known for many years (Laughlin et al., 1981, Levine and Simpser 1982). The factors involved in this are listed in Table 3. Infants and children with Down Syndrome have the mid face hypoplasia with small nares and reduction in the size of the pharynx in association with a large tongue which has a tendency to prolapse backwards during sleep.

The upper airway is frequently small and there are often increased secretions. Tonsillar hypertrophy also plays a role. (Phillips et al., 1988) Other factors known to cause alveolar hypoventilation in other patients are obesity and hypotonia, and these are well known features of Down Syndrome. Atlantoaxial instability can cause acute airway obstruction with hyperextension of the neck during sleep, anesthesia, and sports related activities. This can be life-threatening (Pueschel, 1988).

The high incidence of obstructive sleep apnea in Down Syndrome has been demonstrated by Marcus et al., 1991. Alveolar hypoventilation results in hypoxemia, elegantly demonstrated by Southall et al. (1987), hypocapnia and respiratory acidosis. These three factors act independently on the pulmonary arterioles to produce vasoconstriction (Malik and Kidd 1973, a and b). Vasoconstriction causes pulmonary hypertension and increase in sheer stress in the blood vessel walls, and the tendency to progress to pulmonary vascular disease. (Table 4)

Things that should alert the clinician to the possibility of obstructive sleep apnea in Down Syndrome are excessive and loud snoring with disturbed sleep and

**Table 3.** Factors Predisposing to Alveolar Hypoventilation in Children With Down Syndrome

MID FACE HYPOPLASIA
GLOSSOPTOSIS
SMALL UPPER AIRWAY
TONSILS and ADENOIDS
INCREASED SECRETIONS
OBESITY
HYPOTONIA
ATLANTO-AXIAL INSTABILITY
CONGENITAL HEART DISEASE

**Table 4.** Mechanisms of Action of Alveolar Hypoventilation

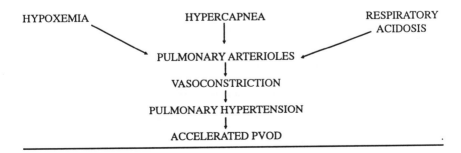

bed wetting (Spicer 1988; Silverman 1988), difficulty in wakening and somnolence during the day, behavioral problems and poor growth (Marcus et al., 1991). (Table 5) Some patients with marked obstructive sleep apnea will need surgical intervention. Tonsillectomy and adenoidectomy has not been universally successful, but the procedure described by Dr. Marshall Strome elsewhere in this book where a comprehensive procedure for enlargement of the pharynx and reduction of the tongue mass together with tonsillectomy has great promise. Early tracheostomy may be required. If there is major atlantoaxial dislocation, stabilization of this joint may be life saving.

## RECOMMENDATION FOR MANAGEMENT OF HEART DISEASE IN DOWN SYNDROME
### Echocardiogram

Since the incidence of heart disease in Down Syndrome is so high, and the existence of the heart disease may be masked, all children with Down Syndrome should have an echocardiogram in the first 2 months of life.

### Trivial Heart Disease

If the heart disease is trivial, such as a very small ventricular septal defect, then intervention is not indicated, and prophylaxis for bacterial endocarditis and follow-up is all that is required.

### Major Heart Disease

If the heart disease is major, as with atrioventricular canal defect, or large ventricular septal defect, and the defect is recognized early, by fetal echocardiogram or shortly after birth, the cardiac catheterization should be planned for around 3 months of age, with corrective surgery to follow shortly thereafter. If the infant is seen in congestive heart failure, anti-congestive measures should be started and an accelerated course of action, catheterization, and surgery set in motion.

In the event that surgery is delayed, it is likely that irreversible pulmonary vascular obstructive disease will have become established, and in these patients

Table 5.  Obstructive Sleep Apnea in Down Syndrome

| ALERTING SIGN | |
|---|---|
| NIGHT | SNORING |
| | DISTURBED SLEEP |
| | BED WETTING |
| DAY | DIFFICULTY IN WAKING |
| | SOMNOLENCE |
| | BEHAVIORAL PROBLEMS |
| | POOR GROWTH (DOWN GROWTH CURVES) |

surgical intervention is contraindicated, and long-term management of the chronic heart disease planned for. As these patients become older, they become cyanotic and venesection may be necessary to maintain the hematocrit around 60%—a level which combines optimal oxygen delivery, and minimal increase in blood viscosity.

For children with atrioventricular canal the repair comprises a dacron patch closure of the ventricular septal defect, and repair of the cleft in the mitral valve, and a pericardial patch to the primum atrial septal defect (Rastelli et al., 1968). Repair of large ventricular septal defects with a dacron patch is similarly urgent in the first 9 months of life. Isolated atrial septal defects and small ductuses are less urgent matters. The timing of tetralogy of Fallot is often dictated by the extent of the right ventricular outflow tract obstruction and the depth of the cyanosis.

## THE ROLE OF SURGICAL INTERVENTION FOR PATIENTS WITH DOWN SYNDROME WHO HAVE HEART DISEASE

Actuarial survival tables of patients with heart disease without intervention are impossible to achieve nowadays, but Fabia and Drollette (1970) suggested that mortality might be as much as 40% by 10 years of age. Since surgical repair for atrioventricular canal was introduced from the Mayo Clinic in 1968, the mortality rate for repair of this lesion during the 1980s from 10 reported series, has varied between 0 and 33% (Bull et al., 1985). In data from Toronto, Freedom et al., 1989, reporting on children who had surgery up to 1986, revealed 21% early deaths and 4.5% late deaths. Even more recently, Pozzi et al. (1991) from Atlanta reported a 5.7% early death rate with 8.5% late deaths. The deaths occurred mostly in children with Down Syndrome and progressive vascular obstructive disease. Although Morray et al. (1986) suggested an increased surgical risk in children with Down Syndrome, Vet and Otterkamp (1989) demonstrated that at least for atrioventricular canal defects, survival rate was much better in the Down Syndrome children.

Thus, with current surgical management, deep hypothermic surgical arrest, precise repair of the defects, and obsessional care of the patient in the postoperative period, the surgical mortality in children without fixed progressive pulmonary vascular obstructive disease may be expected to be low.

## LIFE AFTER SURGERY

It is very rare to have a normal "heart" following successful repair of atrioventricular canal defect. While the ventricular septal defect can successfully be closed using a Dacron patch, and the atrial septal defect using a pericardial patch, the surgical technique of using as much of the endocardial cushion tissue as possible to create a competent mitral valve is not always successful. While in most cases the anterior and posterior parts of the common AV valve can be seen together to form a competent anterior leaflet of the mitral valve, mitral regurgitation is fairly common, but is often mild. Conduction defects, such as complete heart block are also now rare. As mentioned before, pulmonary vascular obstructive disease, if it is present, will persist following surgical repair. For these reasons, surgery is recommended before the 6th month of life.

The long-term listing of the children now growing up following timely and competent repair of their heart defects is now unfolding, and it is not possible to foretell how they will do in the long term. Mitral valve prolapse is common in older patients with Down Syndrome, and this may lead to difficulty. However, many patients are now in their late teens and are doing well.

It is likely that timely surgical repair will prevent the premature mortality in their 3rd and 4th decades of those who developed the Eisenmenger syndrome.

The outlook therefore, is bright for children with significant heart disease and Down Syndrome. Vigilance in infancy, and early surgical repair hold great promise.

## REFERENCES

Bull C, Rigby ML, Shinebourne EA (1985). Should Management of Complete Atrioventricular Canal Defects be Influenced by Coexistant Down Syndrome? Lance 1:1147-49.

Chi TL, Krovetz J (1975). The Pulmonary Vascular Bed in Children with Down Syndrome. J Peds 86:533-538.

Clapp S, Perry BL, Farooki ZQ, Jackson WL, Karpawich PP, Hakimi M, Arciniegas E, Green EW, Pinsky WW (1990). Down's Syndrome, Complete Atrioventricular Canal, and Pulmonary Vascular Obstructive Disease. J Thorac Cardiovasc Surg 100:115-21.

Clark EB, Takao A (1990). Overview: A Focus for Research in Cardiovascular Development. In: Clark EB, Takao A (eds.), "Developmental Cardiology, Morphogenesis and Function." Mt. Kisco, NY: Futura Publishing Co. pp. 3-12.

Cooney TP, Thrulbeck WM (1982). Pulmonary Hypoplasia in Down's Syndrome. N Eng J Med 307:1170-1173.

DuShane JW, Krongrad E, Ritter DG, McGoon DC (1976). The Fate of Raised Pulmonary Vascular Resistance after Surgery in Ventricular Septal Defect. In: Kidd L, Rowe RD (eds), The Child with Congenital Heart Disease after Surgery. Mount Kisco, N.Y.: Futura Publishing Co. pp. 299-312

Eisenmenger V (1897). Die Angieborenen defect die Kammerschiedewande des Herzens. Ztschr. f klin Med 32 Suppl 1.

Fabia J, Drollette M (1970). Life Table up to Age 10 for Mongols with and without Congenital Heart Defect. J Ment Def Res 14:235-242.

Ferencz C, Rubin JD, McCarter RJ, Boughman JA, Wilson PD, Brenner JI, Neill CA, Perry LW, Hepner SI, Downing JW (1987). Cardiac and Noncardiac Malformations: Observations in a Population-Based Study. Tetralogy 35:367-378.

Freedom RM, Rebeyka I, Smallhorn JF, Rabinovitch M, Musewe N, Turner-Gomez S, Alexander D, Thompson L, Williams WG, Coles J, Trusler GA (1989). Late Postoperative Functional Results after Complete Repair in Infancy for Atrioventricular Septal Defect. Perspectives in Ped Card Surg. Vol II. 1:135-138.

Friedli B, Kidd BSL, Mustard WT and Keith JD (1974): Ventricular Septal Defect with Increased Pulmonary Vascular Resistance. Late Results of Surgical Closure. Am J Cardiol.

Greenwood RD, Nadas AS (1976). The Clinical Course of Cardiac Disease in Down's Syndrome. Pediatrics 58:893-897.

Loughlin GM, Wynne JW, Victorica BG (1981). Sleep Apnea asa Possible Cause of Pulmonary Hypertension in Down Syndrome. J Pediatr. 98:435-437.

Levine OR, Simpser M (1982). Alveolar Hypoventilation and Cor Pulmonale Associated with Chronic Airway Obstruction in Infants with Down Syndrome. Clin Pediatr 21:25-29.

Malik A and Kidd BSL: Time Course of the Pulmonary Vascular Resistance to Hypoxia in Dogs. Am J Physiol. 224: 1, 1973.

Malik AB and Kidd BSL: Independent Effects of Changes in $H^+$ and $CO_2$ Concentrations on Hypoxic Pulmonary Vasoconstriction. J Appl Physiol. 34:318, 1973.

Marcus CL, Keens TG, Bautista DB, von Pechmann WS, Davidson Ward SL (1991). Obstructive Sleep Apnea in Children with Down Syndrome. (Pediatrics - In Press).

Marino B, Vairo U, Corno A, Nava S, Guccione P, Calabro R, Marcelletti C (1990). Atrioventricular Canal in Down Syndrome. AJDC 144:1120-1122.

Marino B, Papa M, Guccione P, Corno A, Marasini M, Calabro R (1990). Ventricular Septal Defect in Down Syndrome. AJDC 144:544-545.

Morray JP, MacGillivray R, Duker G (1986). Increased Periopeative Risk Following Repair of Congenital Heart Disease in Down's Syndrome. Anesthesiology 65:221-224.

Park SC, Matthews RA, Zuberbuhler JR, Rowe RD, Neches WH, Lenox CC (1977). Down Syndrome with Congenital Heart Malformation. 131:29-33.

Phillips DE, Rogers JH (1988). Down's Syndrome with Lingual Tonsil Hypertrophy Producing Sleep Apnoea. J Laryng and Oto 102:1054-1055.

Pozzi M, Remig J, Fimmers R, and Urban AE (1991). Atrioventricular Septal Defects: Analysis of Short- and Medium-Term Results. J Thorac Cardiovasc Surg. 101:138-42.

Pueschel SM (1988). Atlantoaxial Instability and Down Syndrome. Pediatrics 81:879-880

Rastelli GC, Ongley PA, Kirklin JW, McGoon DC (1968). Surgical Repair of the Complete Form of Persistent Common Atrioventricular Canal. J Thorac Cardiovasc Surg. 55:299-308.

Schneider DS, Zahka KG, Clark EB, Neill CA (1989). Patterns of Cardiac Care in Infants with Down Syndrome. AJDC 143:363-365.

Silverman M (1988). Airway Obstruction and Sleep Disruption in Down's Syndrome. BMJ 296:1618-19.

Spicer RL (1984). Cardiovascular Disease in Down Syndrome. Peds Clinics of N Amer 31:6, pp 1331-1343.

Southall DP, Stebbins VA, Misza R, Long MH, Croft CB, Shinebourne EA (1987). Upper Airway Obstruction in the Hypoxemia and Sleep Disruption in Down Syndrome. Dev. Med and Child Neurol. 29:734-742.

Stradling JR (1988). Sleep Disruption in Down's Syndrome. BMJ 297:289-90.

Vet TW, Ottenkamp J (1989). Correction of Atrioventricular Septal Defect: Results Influenced by Down Syndrome? AJDC 143:1361-1365.

Wright TC, Orkin RW, Destrempes M, Kurnit DM (1984). Increased Adhesiveness of Down Syndrome Fetal Fibroblasts in vitro. Proc Natl Acad Sci USA 81:2426-2430.

Yamaki S, Horiuchi T, Sekino Y (1983). Quanitative Analysis of Pulmonary Vascular Diseae in Simple Cardiac Anomalies with the Down Syndrome. Am J Cardiol 51:1502-06.

Yamaki S, Horiuchi T, Takahashi T (1985). Pulmonary Changes in Congenital Heart Disease with Down's Syndrome: Their Significance as a Cause of Postoperative Respiratory Failure. Thorax 40:380-386.

# Endocrine Function in Down Syndrome

Ernest E. McCoy, MD

In case reports nearly every type of endocrine disorder has been described in Down syndrome subjects. Most of the published work, however, deals with three areas of endocrine function. By far the most prominent is that of thyroid dysfunction. With the changes in lifestyle of adult Down syndrome subjects from institutional to community living, gonadal function is becoming of increased importance. Short stature has long been recognized as one of the physical characteristics of Down syndrome. Recent studies have explored the mechanism of short stature in Down syndrome and raise questions as to the efficacy of treatment with growth hormone. These three areas of endocrine function in Down syndrome will be examined in this paper.

## THYROID DISEASE IN DOWN SYNDROME

Down syndrome subjects have a higher incidence of thyroid disease in all age ranges compared to non Down syndrome control population.

### Incidence of Hypothyroidism in Down Syndrome in a Congenital Hypothyroidism Screening Program

Fort et al. (1984) examined the results of screening for congenital hypothyroidism in 945,000 newborns screened in New York State over a 3-9/12 year period. In this population there were 1130 infants with the diagnosis of Down syndrome. There were 250 normal newborns in whom congenital hypothyroidism was detected. The incidence in this population as shown in Table 1 was 1:3800. Among the 1130 newborns with Down syndrome 12 cases of hypothyroidism were detected, of which 8 had persistent hypothyroidism. The incidence of congenital hypothyroidism in Down syndrome was 1:141 or 27 times as frequent as in non Down syndrome newborns.

The cause of this increased frequency of hypothyroidism in Down syndrome newborns is unknown. It was the recommendation of Fort et al. that if TSH is elevated and serum thyroxine is normal, that supplemental thyroxine be given

**Table 1. Incidence of Hypothyroidism in Down Syndrome (D.S.) Within a Neonatal Thyroid Screening Program in New York State: 1978-1982**

| | |
|---|---|
| Total Newborns Screened | 945,000 |
| Congenital Hypothyroidism Detected | 250 |
| Incidence Congenital Hypothyroidism in Normal Children | 1:3800 |
| Newborn D.S. Infants Screened for Congenital Hypothyroidism | 1,130 |
| D.S.- Detected with Congenital Hypothyroidism | 12 |
| - with Persistent Hypothyroidism | 8 |
| Incidence Congenital Hypothyroidism in D.S. | 1:141 |

(Adapted from P. Fort et al. J Ped 104:545:1984)

until age 3 years to avoid thyroid decompensation during the critical period of brain growth. Some endocrinologists would favor careful monitoring rather than starting thyroxine in this age group.

## Thyroid Function in Young Children With Down Syndrome

Cutler et al. (1986) studied thyroid function in 49 Down syndrome children aged 4 months to 3 years and 49 age matched, developmentally delayed controls, followed in a Developmental Evaluation Clinic.

Of the 49 Down syndrome children, 3 had congenital hypothyroidism for which they received thyroxine (Table 2). Two additional children had acquired disease - 1 had autoimmune hypothyroidism and 1 hyperthyroidism. Both children improved with appropriate therapy. Of the remaining 44 children, 13 (26%) had compensated hypothyroidism characterized by elevated TSH and normal serum T4 levels. When 10 of these 13 children were retested, in 8 the TSH and T4 levels had returned to normal, age matched values. None of the 49 age matched controls had abnormalities of their T4 or TSH levels. In the group of Down syndrome children with elevated TSH and normal T4 levels, none required treatment and growth patterns were normal.

The study showed a high incidence of abnormalities of thyroid function in this group of Down syndrome children. The study pointed out the need for periodic thyroid function testing of Down syndrome children. The cause of the transitory increase in TSH values is unknown. The fact that in a number of cases the TSH values returned to normal would lead to caution in initiating thyroid replacement therapy without careful serial evaluation of clinical function as well as laboratory values.

**Table 2.  Thyroid Function in Young Down Syndrome (D.S.) Children**

| Study Subjects | |
| --- | --- |
| | 49 D.S. children age 4-36 months<br>49 age matched children with<br>developmental delay |
| D.S. with Cong. Hypothyroidism | 3 |
| D.S. with Acquired Thyroid<br>Disease | 2 (1 Hypo - 1 Hyperthyroid) |
| D.S.; Compensated Hypothyroidism<br>(Increased TSH; Normal Serum T4)<br>Growth Parameters Normal | 13 (26%) |

(Adapted from Cutler et al. Amer J Dis Child 140:479:1986)

## Thyroid Dysfunction in Children and Adults with Down Syndrome

Friedman et al. (1989) checked for evidence of thyroid dysfunction in a group of 138 Down syndrome subjects living in the community who were seen for routine health care services. As shown in Table 3, twenty-eight (20.3%) of the 138 had previously unrecognized hypothyroidism. Of these 28, 11 had non-compensated hypothyroidism characterized by low serum T4 and elevated TSH values. Seventeen of the 28 Down syndrome subjects had compensated hypothyroidism characterized by an elevated serum TSH and normal T4 level. An additional 2 patients had unrecognized hyperthyroidism. Sixty-six of the 138 subjects were tested for the presence of thyroid autoantibodies. Of these 66 subjects, 26 or 39% tested positive: 21 had antimicrosomal antibodies, 2 antithyroglobulin anti-

**Table 3.  Thyroid Dysfunction in Down Syndrome**

| | |
| --- | --- |
| Patients: | 138 - Community Living. Ages 2-59 years |
| Hypothyroid: | 28 not previously recognized (20.3%)<br>(11 with low serum T4 - high TSH)<br>(17 with normal T4 - high TSH) |
| Hyperthyroid: | 2 previously not recognized |
| Thyroid: | 66 of 138 tested. 26 (39%) positive |
| Antibodies: | Antimicrosomal antibodies - 21<br>Antithyroglobulin antibodies - 2<br>Both types of antibodies present - 3 |

(Adapted from D. Freidman et al. Arch Int Med 149:1990:1990)

bodies and in 3 subjects both types of antibodies were present. Twenty of the 28 patients with hypothyroidism or low thyroid reserve were tested for thyroid antibodies. Nineteen of the 20 or 95% had thyroid antibodies present. This paper showed the presence of noncompensated and compensated hypothyroidism over the age range of 11 to 60 years. Because of the high incidence and wide age range in which thyroid disease occurs, continued periodic checks of thyroid function are necessary in the Down syndrome population, regardless of age.

The high incidence of thyroid disorders in Down syndrome in wide geographical areas has been noted by other investigators. Dinani and Carpenter (1990) in England noted a 40% incidence of thyroid disease in 106 adult Down syndrome subjects. Baxter et al. (1975) in Australia found abnormal thyroid function in 7 of 11 middle aged or elderly Down syndrome subjects. Pozzan et al. (1990) in Italy and Lejeune et al. (1990) in France also found an increased incidence of thyroid disease in children and adults with Down syndrome.

The report of Lejeune is of interest for the low levels of 3, 3', 5' triiodothyronine (RT3) in a group of Down syndrome children - $340 \pm 147.7$ pmol/L compared to $408.56 \pm 112.1$ pmol/L in normal, age matched controls. The biochemical and physiological significance of the finding is not known, but warrants further study.

Sharav et al. (1988) studied 147 patients with Down syndrome to determine if there was a relationship between elevated TSH and growth delay. Infants with elevated TSH values had delayed growth parameters for head circumference and height and weight compared to normal infants. When comparison was made between Down syndrome subjects with high versus normal TSH values for 5th to 95th percentile head circumference, there was overlap with Down syndrome subjects having normal TSH values below the 5% for head circumference, while subjects with high TSH values had head circumferences within the 5th to 95th percentile. However the head circumference of more Down syndrome children with high serum TSH values fell below the 5th percentile than those with normal TSH values. The same trend was found for the parameters of height and weight in those Down syndrome children with elevated TSH values. Delayed growth was not found in the study of Cutler (1986), although sample size was smaller. The study of Sharev does show the necessity of continued health supervision of young Down syndrome children to ensure their growth parameters are proceeding in a consistent manner and that thyroid function is assessed at period intervals.

## GONADAL FUNCTION IN DOWN SYNDROME SUBJECTS

Considerable progress has been made recently in understanding gonadal function in young adult subjects with Down syndrome. It has been known for years that women with Down syndrome were fertile but until recently there had not been a case where a Down syndrome male was documented as having fathered a child. This section will review recent studies of gonadal function in Down syndrome male and female subjects.

Hsiang et al. (1987) at John Hopkins Hospital evaluated the gonadal function in 100 home-raised Down syndrome subjects comprised of 53 boys and men and 47 girls and women. Their study showed the early onset of gonadal dysfunction in some Down syndrome children. In a group of 25 prepubertal Down syndrome children, 18 of 25 had TSH levels below or within the normal range for age. However, 3 boys and 4 girls had significantly elevated serum TSH levels. Three boys and 1 girl had elevated serum LH levels. However, in both men and women in this study the age for the onset of puberty and completion of puberty was within the normal range.

The stretched penile length and testicular volume was determined in 33 sexually mature men with Down syndrome. As shown in Figure 1, the mean stretched penile length (11.6 cm) was significantly less (p) than the mean for normal men (13.3 cm). Testicular volume in Down syndrome men was determined using the formula Volume = $\pi/6$ x length x width$^2$. The mean testicular volume was 11.8 ml in Down syndrome men which was significantly less (p) than the 18.6 ml reported for normal men.

Hsiang et al. (1987) determined serum LH and FSH levels in this same group of men. As shown in Figure 2 they found the mean level of both LH and FSH to be increased. The mean serum level of FSH was 18.7 mIU/ml, significantly increased (p) compared to the mean for normal men of 7.2 mIU/ml. The mean level of LH was also increased (16.9 mIU/ml in Down syndrome men compared to the mean level of normal men 9.1 mIU/ml (p). In contrast, the level of serum testosterone was similar between Down syndrome and normal men. The above

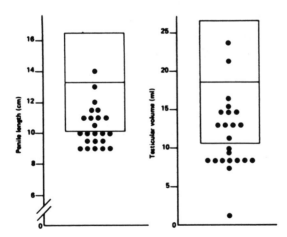

Fig. 1.    Stretched penile length and testicular size were measured in 23 sexually mature men with Down Syndrome. Testicular volume was determined by applying the formula: volume = $\pi/6$ x length x width$^2$. The boxed area represents the mean ± 2 SD for penile length [Lee et al., 1980] and for testicular volume [Zachmann et al., 1974] in normal men (Hsaing et al., 1987, with permission).

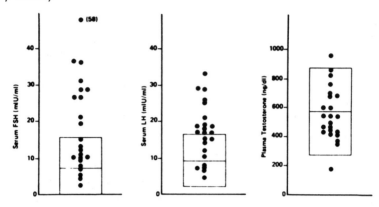

Fig. 2. Serum levels of follicle-stimulating hormone (FSH) and luteinizing hormone (LH) and plasma levels of testosterone were determined in 23 sexually mature men with Down Syndrome. The boxed areas represent the mean ± 2 SD for normal adult men (Hsiang et al., 1987, with permission).

findings are in keeping with an increased incidence of primary gonadal dysfunction in Down syndrome men.

Gonadal function was assessed by LH and FSH levels in 14 sexually mature women with Down syndrome by Hsiang et al. (1987). As shown in Table 4 adapted from this work, the mean level of FSH was 25.4 mIU/ml and for LH 25.4 mIU/ml which was higher than the mean values for normal women in the follicular phase of the menstrual cycle of 8.7 ± 3.3 mIU/ml for serum FSH and 7.7 ± 4.3 mIU/ml for serum LH values. This is suggestive of a higher incidence of primary gonadal dysfunction in mature women with Down syndrome compared to normal women.

Goldstein (1988) studied aspects of menarche, menstruation and sexual relations in 15 adolescent Danish girls with Down syndrome and compared these

**Table 4. Serum LH and FSH Levels in Sexually Mature Down Syndrome and Normal Women**

|  | FSH (mIU/ml) | LH (mIU/l) |
|---|---|---|
| - Mean levels for 14 D.S. adult women at varying stage menstrual cycle | 24.4 | 25.4 |
| - Normal levels for adult women (±1 S.D.) during follicular phase | 8.7±3.3 | 7.7±4.3 |

(Adapted from Hsiang et al. Amer J Med Genet 27:499-1987)

findings to 33 age matched female controls living in the same county. As shown in Table 5, the age of menarche was the same in both groups -13.6 years in the Down syndrome girls and 13.5 years in the control group. The average length of the menstrual cycle was also similar being 28.6 days for controls and 28.3 days for Down syndrome girls. The average length of bleeding during the cycle was 5.5 days for Down syndrome and 5.4 days for controls. Sexual relations were compared in the two groups. Thirteen of the 15 Down syndrome girls for which information was available had not had sexual intercourse, compared to 24 of 33 controls who had a positive history of sexual relations. This study did not determine if ovulation occurred during the menstrual cycle, but because of the regularity of vaginal bleeding the author felt most of the subjects did ovulate.

The study did conclude that menarche, length of menstrual cycle and vaginal bleeding in this group of young Down syndrome women was not different from that of healthy, normal women of the same age group.

Until recently it was thought that males with Down syndrome were sterile as there was no documented case where a man with Down syndrome had fathered a child. This widely held belief was based on reports of decreased spermatozoa counts and of increased hyalinization of seminiferous tubules in Down syndrome adults. This illusion was shattered with the paper by Sheridan et al. (1989) which has provided unequivocal data that a Down syndrome male fathered a pregnancy in a non Down syndrome mentally retarded female. In this case report, the Down syndrome man had unprotected sex for a period of 3 months with a mentally subnormal girl living in the same local authority house. The couple were seen after 8 weeks of amenorrhea and an ultrasound examination confirmed a singleton pregnancy. After counselling the couple agreed to a chorion villus biopsy. The

Table 5.  Menarche, Menstruation and Contraception in a Danish County Down Syndrome (D.S.) Cohort

| | | |
|---|---|---|
| Study population: | 15 D.S. females average age 18.2 years 33 Normal females average age 18.1 years | |
| Average Age Menarche: | D.S. - 13.6 years | Controls - 13.5 years |
| Average Length of Menstrual Cycle | D.S. - 28.3 days | Controls - 28.6 days |
| Average Duration Bleeding in Cycle | D.S. - 5.5 days | Controls - 5.4 days |
| Sexual Contact | D.S. - 0/13 positive - 2 no information | Controls - 24/33 positive - 9 no contact |
| Contraception Use | D.S. - 4/15 | Controls - 22/33 |

(Adapted from H. Goldstein Eur J Obst Gyn Reprod Biol 27:343:1988)

procedure and immediate post biopsy course was uneventful. Despite counselling to cease intercourse, the couple continued to have sexual relations. This resulted in bleeding 2 weeks post-biopsy and a spontaneous loss of the pregnancy several weeks later. Cytogenetic and molecular genetic techniques were used to demonstrate that the Down syndrome man had fathered the child. Chromosome analysis showed the chromosome complement of the putative father to be 47xy+21 in lymphocyte metaphases. The woman had a normal 46xx chromosome complement. Chromosome studies from the chorionic villus biopsy showed the fetus to be a 46xy male. Paternity was confirmed by analysis of DNA fingerprint patterns obtained from the fetal chorionic villus sample and from both parents. The DNA fingerprint pattern (Table 6) showed 53 bands in the mother and the same number in the putative father. The fetus had on analysis 55 bands, 32 of which were shared by the mother. There were 22 bands that were shared exclusively with the reported father. Although other theoretical possibilities for paternity were considered, they were so unlikely that it was concluded that the Down syndrome man was the father of the pregnancy.

This report using present day molecular biology techniques has shown beyond reasonable doubt that a Down syndrome man fathered a pregnancy. This has important implications for counselling, as prior to this paper the common advice given was that there were no instances where a Down syndrome man had fathered a child. With more open living conditions for Down syndrome subjects than have existed in the past, advice to the care givers will need to include the possibility that the Down syndrome males under their care may be fertile. Professor Babrow (personal communication) has shown that the same Down syndrome man fathered a second pregnancy that resulted in the birth of a normal boy.

## GROWTH PATTERNS IN DOWN SYNDROME

One of the long recognized physical characteristics of Down syndrome subjects is short stature. This decrease in height is present at all life stages. Clementi et al. (1990) undertook measurements of 688 consecutive newborn infants with Down syndrome and compared them to 6890 normal newborn infants in the region of North East Italy. Measurements of length, weight and head circumference were done on newborns from 33 to 43 weeks gestation. Growth curves from

Table 6. DNA Fingerprint Bands from a Down Syndrome Father Non Down Syndrome Mother and Fetus

| | |
|---|---|
| Total number of DNA bands in father or mother | 53 |
| Total number of DNA bands in fetus | 55 |
| DNA bands shared by mother and fetus | 32 |
| DNA bands shared exclusively by father and fetus | 22 |

(Adapted from R. Sheridan et al. J Med Genet 26:294-298:1989)

the 10th to the 90th percentiles were constructed for the control group from 32 to 43 weeks gestation at birth. Although most Down syndrome infants fell between the 10th and 90th percentile for normal infants in respect to head circumference, height and weight, overall they were lower than that of the controls. There were a moderate number of Down syndrome newborns that fell far below the 10th percentile for normal newborns in all 3 parameters. This suggests that growth in Down syndrome is prenatally reduced.

Growth charts have been constructed for Down syndrome children to ages 14 (Piro 1990) or 18 (Cronk 1988) years of age. The study of Cronk et al. (1988) found there was a 20% reduction in growth rate in infancy (3 to 36 months) in both sexes, only a 5% reduction in growth rate of girls between 3 to 10 years and about a 10% reduction in boys between 3 and 12 years of age. A more marked reduction in growth rates occurred in adolescent years, ie. a 27% decrease in girls age 10 to 17 years and a 50% reduction in boys age 12 to 17 years indicating a decreased adolescent growth spurt in Down syndrome. Although both the study of Cronk et al. (1988) and that of Piro et al. (1990) showed a reduction in growth rates at all ages, there has been a progressive improvement in growth of Down syndrome children compared to earlier reports.

There is growth delay in Down syndrome children but how this occurs is unclear. Several mechanisms can be proposed but none have been shown to have a definite role. Chromosome imbalance due to the additional chromosome 21 may per se interfere with cellular metabolism to a sufficient degree to influence growth. The loss of the short arm of the X chromosome has been put forth to account for the short stature in Turners syndrome (Ferguson-Smith 1965) where a loss of an X chromosome occurs. Abnormalities in the cellular content and transport of sodium and potassium and in intracellular pH have been found in Down syndrome cells (McCoy 1980, 86, 88). Whether these can affect growth rates is unclear. Recently it has been proposed that decreased growth rates in Down syndrome infants may be due to growth hormone deficiency secondary to hypothalamic dysfunction (Castells 1991). Further studies will need to be done to substantiate this proposal.

There is considerable controversy as to the efficacy of administration of human growth hormone to short children with normal increases of growth hormone on provocative testing with agents such as L-DOPA or arginine, or physiological tests such as exercise or sleep. Anneren et al. (1986) showed that the concentrations of plasma growth hormone during sleep or insulin-arginine stimulated release in 5 Down syndrome children ages 3-½ to 6-½ years (Table 7) was similar to that seen in normal children.

Growth hormone is thought to exert its major action by stimulation of IGF1 production. Sara et al. (1983) showed that the radioimmunoassay level of IGF1 (Som.A) was increased in infancy in Down syndrome compared to normal children, but failed to rise with increasing age as occurs in the normal population. The radioreceptor level of Somatomedin A (IGF1) was increased at all ages in the

**Table 7.   Plasma Growth Hormone Levels (g/l) Following Sleep or Insulin Argenine Stimulated Release in 5 Down Syndrome Children**

|  | Maximum G.H. - sleep - | Maximum G.H. Insulin-Argenine |
|---|---|---|
| Case 1 | 55.0 | 16.7 |
| Case 2 | 16.7 | 13.7 |
| Case 3 | 5.2 | 19.9 |
| Case 4 | 8.8 | 5.4 |
| Case 5 | 9.6 | 20.0 |

(Adapted from G. Annerén et al. Arch Dis Child 61:48-52:1986)

Down syndrome subjects tested. It was thought that the form of somatomedin elevated, was ineffective as a skeletal growth factor. Thus, the low level of the radioimmunoassayable form of IGF1 during childhood and adulthood plus an inactive form of radioreceptor assayed IGF1 might account for the growth delay in Down syndrome. Annerén et al. (1990) in later studies showed that the radioimmunoassay form of IGF1 remained low in Down syndrome children and adults but increased to the normal range when human growth hormone was administered to 5 Down syndrome children for a period of 6 months. Growth velocity increased in these Down syndrome children during growth hormone administration from 2.3-2.8 cm to 3.3-5.8 cm per 6 months. These studies of Annerén et al. (1986, 1990) showed that Down syndrome children respond in a normal manner to increase growth velocity during growth hormone administration. No deceleration of growth velocity was observed in these Down syndrome children over the 4 years since growth hormone was discontinued.

Non Down syndrome children with normal variant short stature also have low levels of plasma IGF1. On growth hormone administration, they also increase their growth velocity and their plasma IGF1 levels increase to the normal range for age (Rudman 1981). There is insufficient data (Cara 1990) and the results of long term treatment of such children marginal (Wit 1990) to recommend widespread treatment of normal, short stature children with growth hormone. There should be a similar cautious approach to the treatment of short Down syndrome children with growth hormone. A recent abstract by Torrado et al. (1990) reports increased growth rates, somatomedin C levels (IGF1) and head circumference during growth hormone administration to Down syndrome children who on stimulatory tests for growth hormone release had low growth hormone levels. Castells et al. (1991) in a later abstract propose, on the basis of decreased growth hormone levels to L-DOPA and clonidine stimulated growth hormone release but normal growth hormone levels after growth hormone releasing factor (GHRF), that Down syndrome children may have growth hormone deficiency secondary to hypothalamic dysfunction—i.e., a decreased GHRF release.

It cannot be emphasized too strongly that further controlled and pilot studies need to be carried out before treatment with this potent hormone is administered to Down syndrome children. It is unknown whether any improvement in eventual height or mental function would occur from early long term administration of growth hormone. Also unknown is the possibility of long-term, adverse effects of administration of growth hormone to Down syndrome children.

## OTHER ENDOCRINE DISORDERS

### Insulin-Dependent Diabetes Mellitus

From reports cited previously in this chapter, the incidence of autoimmune thyroid disease is increased in Down syndrome adults. It is generally acknowledged that the presence of one autoimmune disease predisposes the individual to the development of other types of autoimmune disorders. Several case reports have described the development of hyperthyroidism and later of Type 1 diabetes (Ruch 1985) or diabetes with the later development of hypothyroidism (Radetti 1986) in Down syndrome subjects. Although there are a number of case reports of diabetes in Down syndrome, there is insufficient data as to whether the incidence is significantly increased. One can conclude that if autoimmune thyroid disease is present, a history of symptoms relating to diabetes should be covered during periodic visits for medical care.

## REFERENCES

Annerén G, Sara VR, Hall K, Tuvemo T (1986). Growth and somatomedin responses to growth hormone in Down's syndrome. Arch Dis Child 61:48-52.

Annerén G, Gustavson K-H, Sara VR, Tuvemo T (1990). Growth retardation in Down syndrome in relation to insulin-like growth factors and growth hormone. Amer J Med Genet Suppl 7:59-62.

Baxter RG, Martin F.I.R, Myles K, Larkins RG, Heyma P, Ryan L (1975). Down syndrome and thyroid function in adults. The Lancet II:794-796.

Cara JF, Johnson AJ (1990). Growth hormone for short stature not due to classic growth hormone deficiency. Ped Clin N Amer 37:1229-1254.

Castells S, Torrado C, Gelato MC (1991). Growth hormone (GH) responses to GH releasing hormone (GH-RH) suggests hypothalamic dysfunction as a cause for growth retardation in Down syndrome. Ped Res 29:75A.

Clementi M, Calzolari E, Turolla L, Volpato S, Tenconi R (1990). Neonatal growth patterns in a population of consecutively born Down syndrome children. Amer J Med Genet Suppl 7:71-74.

Cronk C, Crocker AC, Pueschel SM, Shea AM, Zackai E, Pickens G, Reed RB (1988). Growth charts for children with Down syndrome: 1 month to 18 years of age. Pediatrics 81:102-110.

Cutler AT, Benezra-Obeiter R, Brink SJ (1986). Thyroid function in young children with Down syndrome. Amer J Dis Child 140:479-483.

Dinani S, Carpenter S (1990). Down's syndrome and thyroid disorder. J Mental Def Res 34:187-193.

Ferguson-Smith MA (1965). Karyotype-phenotype correlations in gonadal dysgenesis and their bearing on the pathogenesis of malformation. J Med Genet 2:142-155.

Fort P, Lifshitz F, Bellisario R, Davis J, Lanes R, Pugliesse M, Richman R, Post EM, David R (1984). Abnormalities of thyroid function in infants with Down syndrome. J Ped 104:545-549.

Freidman DL, Kastner T, Pond WS, O'Brien DR (1989). Thyroid dysfunction in individuals with Down syndrome. Arch Int Med 149:1990-1993.

Goldstein H (1988). Menarche, menstruation, sexual relations and contraception of adolescent females with Down syndrome. Eur J Obst Gyn Reprod Biol 27:343-349.

Hsiang Y-HH, Berkovitz GD, Bland GL, Migeon CJ, Warren AC (1987). Gonadal function in patients with Down syndrome. Amer J Med Genet 27:449-458.

Lejeune J, Peeters M, Rethore MO, DeBlois MC, Devaux JP 1990). Down syndrome and 3'3',5'-Triiodothyronine. Amer J Dis Child 144:19.

McCoy EE, Enns L (1990). Potassium uptake by platelets from Down's syndrome and normal subjects. Life Sci 26:603-606.

McCoy EE, Enns L (1986). Current status of neurotransmitter abnormalities in Down syndrome In: The Neurobiology of Down Syndrome. Ed. C.J. Epstein Raven Press New York pp 73-87.

McCoy EE, Strynadka K (1988). Down syndrome platelets, lymphocytes and lymphoblasts have increased intracellular pH. Clin Res 36:219A.

Piro E, Pennino C, Cammarata M, Corsello G, Grenci A, Lo Giudice C, Morabito M, Piccione M, Guiffrè L (1990). Growth charts of Down syndrome in Sicily. Evaluation of 382 children 0-14 years of age. Amer J Med Genet Suppl 7:66-70.

Pozzan GB, Rigon F, Girelli ME, Rubello D, Busnardo B, Baccichetti C (1990). Thyroid function in patients with Down syndrome. Amer J Med Genet Suppl 7:57-58.

Radetti G, Drei F, Betterle C, Mengarda G (1986). Down syndrome, hypothyroidism and insulin-dependent diabetes mellitus. Helv Paed Acta 41:377-380.

Ruch W, Schürmann K, Gordon P, Bürgin-Wolff A, Girard J (1985). Coexistent caeliac disease, Grave's disease and diabetes mellitus Type I in a patient with Down syndrome. Eur J Ped 144:89-90.

Rudman D, Kutner MH, Blackston RD, Cushman RA, Bain RP, Patterson JH (1981). Children with normal variant short stature: treatment with human growth hormone for six months. New Eng J Med 305:123-131.

Sara VR, Gustavson K-H, Annerén G, Hall K, Wetterberg L (1983). Somatomedins in Down syndrome. Biol Psych 18:803-811.

Sharav T, Collins RM, Baab PJ (1988). Growth studies in infants and children with Down's syndrome and elevated levels of thyrotropin. Amer J Dis Child 142:1302-1306.

Sheridan R, Llerena J, Matkins S, Debenham P, Cawood A, Bobrow M (1989). Fertility in a male with trisomy 21. JMed Genet 26:294-298.

Torrado C, Wisniewski K, Castells S (1990). Growth hormone deficiency in Down syndrome: treatment with synthetic human growth hormone. Ped Res 27:87A.

Wit JM, Rikken B, De Munick Keizer-Schrama S.M.R.F, Oostdijk W, Gons M, Otten BJ, Delemarre-Vande Waal HA, Reeser M, Waelkens JJJ (Dutch Growth Hormone Working Group) (1990). Growth hormone therapy for short, slowly growing children without classic growth hormone deficiency: 3 year results. Acta Paed Scand Suppl 370:186.

# Susceptibility to Infectious Disease in Down Syndrome

David J. Lang, MD

It is generally accepted that persons with Down Syndrome (DS) exhibit an increased frequency of infections, especially those involving the head, neck and respiratory tract including sinuses, ears, and airway. Enlarging upon these generalizations, we must distinguish increased susceptibility to infections from an enhanced or more exaggerated response to them. Finally, in order to further investigate, understand, and cope with these episodes, it is useful to further categorize the associated responsible deficiencies and defects into those which can be described as functional and those which are structural.

An impression has been gained by clinicians that children with DS experience marked differences from the norm in the frequency of infections. Stanley Levin (1987) who participated in a prior NDSS Symposium commented that infectious diseases in persons with DS occurred 52 times more frequently than among children in matched non-DS populations. In other sources the increase in respiratory infections is placed even higher. Some of these data were derived many years ago in an era when antibiotics were unavailable or of limited scope. In addition, the changes in management of congenital heart disease have markedly improved the cardio-respiratory status of many children with DS and this has certainly favorably influenced the occurrence of respiratory infections. Oster and colleagues (1975) remarked in that infectious diseases occurred during the decade from 1960-1971 in persons with DS 12-fold more frequently than among matched non-DS persons.

In our own studies among 190 children with DS followed at the City of Hope between 1984 and 1987, 30% had experienced one or more episodes of pneumonia, while ear infection had been documented in 62%. Dermal folliculitis occurred in 3-4%, while 5 to 6% had experienced fungal infections of skin or nails (Heide et al., 1990).

The increased incidence of otitis media (OM) which occurs in children with DS (Balkany et al., 1978), is an observation well known to parents, pediatricians, family practitioners, and otolaryngologists. This increase in OM can contribute to hearing disorders and thus to the already significant impairment of learning inherent in DS.

In the early 1960s Blumberg and associates described the Australia Antigen, which ultimately proved to be the surface antigen of Hepatitis B. In the course of characterizing the newly recognized antigen, Blumberg screened serum samples from a wide variety of patients and discovered an elevated prevalence of antigen in specimens from patients with hepatitis, lepromatous leprosy, leukemia, those who had been multiply transfused, and (strikingly) in persons with DS (Blumberg et al., 1967; Bayer et al., 1968). In one study 30% of individuals with DS were found to be HBsAg positive. Ultimately it was shown that long-term institution-alization (prevalent when Blumberg's sera were collected) provided circum-stances for exposure to the antigen, but that persons with DS did become HBsAg carriers at an unusually high rate. Initially there was an attempt by Blumberg to relate these findings to the elevated frequency of antigenemia among leukemic patients and somehow in turn to the increased frequency of leukemia among persons with DS. In addition Blumberg remarked on the report by Stoller and Collmann (1966) of the increased incidence of DS births 9 months after increases in hepatitis, and the observation that convalescent serums from patients with hepatitis (type unknown) could increase the finding of chromosomal aberrations in vitro (cell culture). Ultimately however it became clear that the increased prevalence of HBs antigenemia in DS reflected institutional exposure to hepatitis B coupled with a subtle immunologic deficit which enhanced the likelihood of establishing a chronic carrier state (Table 1).

Considerable attention has been devoted to the inter-relationships of hepatitis and DS. Before the epidemiology of hepatitis had been clarified by the identifica-tion of the several infectious agents of hepatitis A, B, C, delta, and non-A non-B non-C hepatitis as a possible exogenous cause of DS had received some attention. Collmann, Krupinski and Stoller (1966) reported an association of increased births of infants with DS occurring nine months after increases in the annual occurrence of hepatitis in metropolitan Melbourne Australia. Other investigators were unable to confirm this relationship (Stark and Faumeni, 1966; Ceccareli and Torbidoni, 1967; Kashgarian et al., 1970) although again it should be emphasized that, unbeknownst to these investigators, all of these studies were evaluating the impact of various very distinct forms of hepatitis in association with the later birth of a child with DS. As a by-product of these epidemiologic investigations it was noted that the apparent association was more consistently present among older mothers and it was speculated that "the ageing ovum might... become more

Table 1.  Hepatitis B and DS

1) Relationship to Etiology
2) Frequency of Carrier-State
3) Possible Implications for Cellular Immunity

susceptible to attack by a virus" (Collmann et al., 1966). The concept was that a virus or another exogenous (perhaps infectious) factor might influence the disjunction of parental chromosomes. Seasonality of births could influence observations and apparent associations with other seasonal events including infections. More recently (Kallen and Masback, 1988) the effects of seasonality and parity upon the births of children with DS in Sweden was examined. It was confirmed that seasonality did exist with the birth of non-DS children. Although no seasonality could be shown for the overall month of the LMP with the birth of children with DS, some seasonality could be discerned if the DS cases were divided into groups by parity. It was concluded that there may be heterogeneity in etiological factors of importance for DS, where the factor of subfertility may be more common among primiparous women than among multiparous women. Normal birth seasonality may apply to women of higher parity who give birth to children with DS, but not to those of parity 1. They point out cogently that such an explanation for any seasonality detected is more likely than the presence of an exogenous agent since the critical event has, in many cases, taken place in the maternal first meiotic division and thus sometime (perhaps a long time) prior to conception.

Of more relevance today is the apparent increased risk of a carrier state among individuals with DS infected with HBV. This increased risk is similar to that observed among persons with defined impairments of cellular immunity such as among newborn infants or those who are immunosuppressed by disease or by therapeutic interventions. Thus this observation of an increased risk for HBs antigenemia among persons with DS implies the existence of defective cellular immunity. This in turn may be an important factor in the increased susceptibility to other infectious disease in DS.

Table 2 outlines possible general principles any or all of which may be operative to explain the increased risk of infectious disease in DS. We have already made reference to one aspect of a deficient defense mechanism, namely the implied deficiency in cellular immunity.

Other evidence which will be mentioned later suggests the presence in DS of deficient humoral immunity, as well as the existence of other impaired defenses against microbial invasion. Structural abnormalities exist in many persons with DS which may reduce the effective clearance of infectious agents. Finally there are instances where the presence of DS increases the likelihood of exposure to infections. Although frequent early institutional placement of children with DS is largely a thing of the past, the grouping of these children in "infant stim" programs

Table 2.  Possible Factors

---

1) Deficient Defense Mechanisms
2) Abnormal Structural Features
3) Increased Risk of Exposure

---

and certain educational and group residences may increase the likelihood of the transmission of infectious agents. The effects are similar to those experienced in day-care settings for non-DS children, but in some cases with a more prolonged risk for interpersonal transmission of infections dependent upon the level of development among the affected individuals.

Table 3 further categorizes in general terms the deficient defense mechanisms in DS. Spina et al., (1981) found that B cells were markedly reduced in persons with DS. They also showed a decreased T-cell proliferative response to PHA. Raziuddin and Elawad (1990) also demonstrated a reduced proliferative response to concanavalin A. In vitro studies (Epstein and Epstein, 1980; Epstein and Philip, 1987) have documented defective cellular responses to bacterial and viral antigens, and depressed production of cytokines. Others have described a reduction in circulating T-cells and an apparent depressed synthesis of immunoglobulins reflecting a defect in B cell function. Izumi and colleagues (1989) found defective neutrophil chemotaxis among patients with DS and correlated this with the occurrence of periodontal inflammation, disease, and periodontal destruction.

Studies of immunoglobulins and of lymphocytes in DS have shown increased production of immunoglobulins, increased levels of T-suppressor cells and decreased numbers of T-helper cells. Loh and colleagues (1990) have recently described immunoglobulin G subclass deficiencies involving primarily lgG4 and in some cases of lgG2.

Studies of cellular immunologic function in DS have yielded various pertinent data all of which seem consistent with the existence of some functional immune deficit(s). Reduced serum IgM levels, lower white blood cell counts and reduced lymphocyte populations have been found among non-institutionalized DS children (Hann et al.,1979). T-cell immune deficiency in DS has been suggested (Levin et al., 1975; Burgio et al., 1978). However Crossen and Morgan (1980) found no difference between DS persons and normals with reference to mitotic index. Ugazio and coworkers (1978) described what they called "congenital immunodeficiency" in Down Syndrome. Larocca et al., (1988) concluded that there exists a deficient expansion of immature T-cells in DS and thus a reduced capability of the thymocyte pool to differentiate into functionally mature T-cells. These functional disturbances exist in spite of a lack of gross abnormalities of T cell development in the fetal thymus in DS (Cossarizza et al., 1989). Levin (1979) concludes that children with DS have diminished numbers of T-cells as well as functional deficiencies of these cells. Montagna and colleagues (1988) report impairment of allogeneic mixed lymphocyte reactions as well as of NK and

Table 3.  Deficient Defense Mechanisms

---

    1) Lymphocyte Function (cellular immunity)
    2) Immunoglobulins
    3) Autoimmune Disorders

---

NK-like activities (Montagna et al., 1988). Burgio (1983) concludes that Down Syndrome is "a model of immunodeficiency".

The occurrence of so-called autoimmune disorders in patients with DS also clearly indicates the presence of immunological aberrations (Levo and Green, 1977). The excess occurrence of thyroiditis, and certain forms of alopecia, in DS are examples. The increased frequency of leukemia also suggests the influence of one or more of these immunological defects.

Clusters of children with DS exhibiting unusual infections are reported periodically. Cant, Gibson, and West (1987) recently described the more frequent than expected occurrence of bacterial tracheitis in DS. Bacterial tracheitis is usually due to Staphylococcus aureus, although H influenza B and Streptococci have been implicated in other cases. However, it is of interest that an excess of pyogenic infections has not been prominently noted among those with DS. Why this is so is unclear, especially since an excess of intracellular SOD should impact upon and diminish the effectiveness of superoxide production and thus impair cell-killing action. Indeed Cooper and Hall (1988) suggest that the involvement of superoxide radicals in the inflammatory response and the activation of phagocytes might be perturbed by the excess of SOD-1 in DS cell which in turn could lead to increased susceptibility to infection.

Abnormal structural features also contribute to the increased susceptibility to infections among persons with DS (Table 4). The underdevelopment of the maxilla which contributes directly to aspects of the characteristic facial features of DS also is relevant to the increased frequency of nasopharyngeal infectious syndromes and to the increased frequency of OM in DS. As an example, I will focus for a moment upon the eustachian tubes. Congenital anomalies of the eustachian tubes may contribute indirectly and even directly to the existence of chronic OM. Shibahara and Sando (1989) have recently described findings indicating the presence of inadequate and poorly developed eustachian tubes in a child with DS. The normal eustachian tube is well developed with an open lumen and epithelial infoldings. They describe specimens from a child with DS in which the eustachian tube is poorly developed. Where the lumen existed it was found to be collapsed. That the existence of craniofacial abnormalities is associated with OM with effusion is well-known. Table 5 lists some relevant conditions and syn-

**Table 4.  Abnormal Structural Features**

1) Malformation of Skull and Associated Structures:
   a) Mid-Facial Bones Underdeveloped
   b) Sinuses
   c) Middle Ear Abnormalities
   d) Eustachian Tube

2) Hypotonic Musculature

88 / Lang

**Table 5.  Trisomy 21 (Down Syndrome)**

Cleft palate, Micrognathia, Glossoptosis (Pierre Robin Syndrome)
Trisomy 13-15 (Patau Syndrome)
Mandibulofacial Dysostosis (Treacher Collins Syndrome)
Oculoauriculovertebral Dysplasia (Goldenhar Syndrome)
Acrocephalosyndactyly (Apert Syndrome)
Gonadal Dysgenesis (Turner Syndrome)
Craniometaphyseal Dysplasia (Pyle disease)
Osteopetrosis (Albers-Schonberg Disease)
Achondroplasia (Parrot Disease)
Mucopolysaccharidosis (Hunter-Hurler Syndrome)
Orofacial-Digital Syndrome (Mohr Syndrome)
Craniofacial Dysostosis (Crouzon Disease)

(Adapted from Otitis Media in Infants and Children, Bluestone and Klein eds.)

dromes, all of which underscore this important relationship. This table is modified from one by Bluestone and Klein (1988).

Finally the hypotonia, characteristic of DS contributes to the inadequate function of the levator veli palatini muscles, and thus further impairs eustachian tube function. It is clear that a combination of structural and functional changes and deficiencies is crucial to the increased risks for infection among persons with DS.

With this in mind we can turn our attention to various means which have been employed and proposed to modify the DS phenotype and in the process to reduce the risk of infections. Some of these are listed in Table 6.

Various combinations of vitamins, sometimes in huge dosages, have been tried with or without hormonal supplements, (read thyroid for hormone) to modify the DS phenotype and among other things to influence the frequency of infections in DS. Often families are embarrassed to discuss these treatment modalities with their pediatricians or family physicians, fearing that these mainstream doctors will not be sympathetic. But many families will try anything which gives hope, as long as they feel that there is little or no risk to their children. (*Editors' Note*: However, they need to be cautioned that high doses of the fat soluble vitamins C and B6 have had reported significant side effects.) Similar reasoning has led many to embark on courses of "sicca cell therapy" for their children, though thus far there is no

**Table 6.  Attempts to Modify the DS Phenotype**

1) Vitamins
2) Minerals
3) "Cell Therapy"
4) Plastic Surgery
5) Orthodontic Manipulations

clear evidence for benefit beyond the emotional impact gained when a family "pulls together" with the hope and expectation that they may enhance the outlook for their child (Foreman and Ward, 1987; Van Dyke et al., 1990). Whether there exists a significant risk attendant upon the use of "sicca cell therapy" remains unclear. Favorable impact upon infections has not been studied, although unsupported claims for improved immune function and decreased infections have been made (Schmid, 1983). (*Editors' Note*: There are no controlled studies which show any substantive benefit to children with Down Syndrome who have received mega vitamin or sicca cell therapy. Until such studies are available, these forms of therapy cannot be recommended.)

A recent study by Lockitch and co-workers (Lockitch et al., 1989) failed to show an impact upon infection and immunity in DS of orally administered zinc. However prior studies (Kanavin et al., 1988) seemed to indicate that zinc levels were low in DS and it has been suggested that this may be relevant to a defect in immune regulation. In addition it has been claimed that oral zinc supplementation can reduced the frequency of recurrent infections in DS (Franceschi et al., 1988).

It is of interest that selenium supplementation has been reported to increase serum concentrations of lgG2 and lgG4 in children with DS (Anneren et al., 1990). A prior study had documented a reduction in the rates of infection in DS associated with the administration of selenium supplements. It will be of interest to see whether these reports can be confirmed of an effect of exogenous selenium upon immunoregulatory function in DS.

I will conclude by addressing the issues which I believe are those most pertinent to prevention of infectious conditions in children with DS. The cornerstone of preventive therapy is the same as that which we apply to all children, namely the use of immunizations in conjunction, of course, with careful attention to appropriate nutrition, and all other aspects of comprehensive health care. With regard to immunization however I would like to add some specific recommendations beyond the usual ones for non-DS children (Table 7).

In addition to the administration of DTP, attenuated Poliovirus vaccine, HIB and MMR vaccines, I strongly advise HBV immunization for all children with DS. A child with DS who is an HBV carrier may have difficulty enrolling in day-care or even certain schools. Children with DS have been shown to respond to HBV immunization, though with reduced final titers of antibody (Heijtink et al., 1984; Avanzini et al., 1988). I would therefore suggest reimmunization be considered, especially if the individuals live in a high-risk setting. I also suggest the use of the

**Table 7. Prevention**

| | | |
|---|---|---|
| DTP | HBV | |
| Poliovirus | Pneumococcus | |
| HIB | Influenza | |
| MMR | | |

multi-valent pneumococcal vaccine, and I believe physicians should give yearly attention to influenza immunization among all persons with DS. These suggestions do not seem unusually radical given the current recommendations made by the Advisory Committee on Immunization Practices of the Public Health Service that all children receive HBV vaccine (Kolata, 1991). It has also been suggested (Heikkinen et al., 1991) that influenza immunization be adopted as an effective means to prevent otitis media among normal children under certain circumstances.

## SUMMARY

It is generally accepted that persons with Down Syndrome (DS) have increased susceptibility to infection. The existence of deficient host immune function in DS has been suggested. The increased likelihood among individuals with DS of establishing a carrier-state following hepatitis B virus (HBV) infection is a case in point. The nature of the deficient host defence(s) is still not well understood, although various studies have shown both cellular and or humoral immunological defects. Immunoglobulin subclass deficiency has been reported in children with DS and it has been suggested that this contributes to increased susceptibility to infection.

Frequent ear infections among children with DS may be due to anatomical and functional disturbances of the eustachian tubes. Muscular hypotonia may also contribute to eustachian tube dysfunction. Additional chronic and recurrent sino-pulmonary infections which occur frequently in DS may also reflect structural abnormalities connected with the inadequate development of the entire maxillary region, a feature very characteristic of DS and responsible for many facial features associated with trisomy 21. Overall, respiratory infections remain a major problem for persons with DS and, even in the current era of widespread antibiotic usage, these infections contribute significantly to morbidity and mortality in this population. Overall, it appears that a combination of structural and functional disorders underlie the apparent increased risks of infection associated with DS.

Various attempts have been made to modify the DS phenotype, including the occurrence of infections, by administering specific vitamins, minerals, and hormones. Some of these reports are of interest, such as a recent publication documenting increases in the concentration of lgG subtypes in children with DS by supplementation with selenium. This coupled with a suggested reduced rate of infection in children with DS after a 6-month treatment with selenium supplements warrants further study. However most claims of improvement with nutritional supplements have not been confirmed when subjected to rigorous controlled study.

At this time the most significant thing which can be done to protect persons with DS against infections is to provide optimal preventive medical care. The cornerstone of care to prevent infections is the provision of immunization. In addition to the standard recommendations for DTP, poliovirus, HIB, and MMR administration, it is proposed herewith that all persons with DS receive HBV immunization and that consideration be given to the routine provision of pneumo-coccal vaccine and yearly influenza immunization.

## REFERENCES

Collmann RD, Krupinski J, Stoller A (1966). Incidence of infectious hepatitis compared with the incidence of children with Down's syndrome born nine months later to younger and to older mothers. J Mental Defic Res 10:266-268.

Cooper DN, Hall C (1988). Down's syndrome and the molecular biology of chromosome 21. Progr in Neurobiol 30:507-530.

Cossarizza A, Monti D, Montagnani G, et al (1989). Fetal thymic differentiation in Down's syndrome. Thymus 14:163-170

Crossen PE, Morgan WF (1980). Lymphocyte proliferation in Down's syndrome measured by sister chromatid differential staining. Hum Genet 53:311-313.

Epstein LB, Epstein CJ (1980). T lymphocyte function and sensitivity to interferon in trisomy 21 cell. Immunol 51:303-318.

Epstein LB, Philip R (1987). Abnormalities of the immune response to influenza antigen in Down syndrome (trisomy 21) in oncology and immunology of Down syndrome. McCoy EE, Epstein CJ, eds. Alan R Liss, Inc. New York. pp163-182.

Franceschi C, Chiricolo M, Licastro F et al., (1988). Oral zinc supplementation in Down's syndrome: Restoration of thymic endocrine activity and of some immune defects J Ment Defic Research 32:169-181.

Foreman PJ, Ward J (1987). An evaluation of cell therapy in Down syndrome. Aust Paediatr J 23:151-156.

Hann H-W L, Deacon JC, London WT (1979). Lymphocyte surface markers and serum im-munoglobulins in persons with Down's syndrome. Amer J Ment Defic 84:245-251.

Heide F, Lang DJ, Soucek MJ, Van Duyne S, Van Dyke DC, eds (1990). Clinical perspectives in management of Down's syndrome. Springer-Verlag, New York New York.

Heijtink RA, DeJong P, Schalm SW et al., (1984). Hepatitis B vaccination in Down's syndrome and other mentally retarded patients. Hepatology 4:611-614.

Heikkinen T, Ruuskanen, Waris et al., (1991). Influenza vaccination in the prevention of acute otitis media in children. Amer J Dis Child 145:445-448.

Izumi Y, Sugiyama S, Shinozuka D et al., (1989). Defective neutrophil chemotaxix in Down's syndrome; patients and its relationship to periodontal destruction. J Periodontol 60:238-242.

Kallen and Masback (1988). Down's syndrome; seasonality and parity effects. Hereditas 109:21-27.

Kanavin O, Scott H, Fausa O et al., (1988). Immunological studies of patients with Down's syndrome. Acta Med Scan 224:473-477.

Kashgarian M, Rendtorff RC, Fowinkle EW (1970). Epidemics of infectious hepatitis; their relationship to the incidence of Down's syndrome in memphis.J Tenn Med Assoc 63:29-33.

Kolata G (1991). NY Times March 1 pg A16.

Larocca LM, Piantelli M, Valitutti S et al., (1988) Alterations in thymocyte subpopulations in Down's syndrome. Clin Immunol and Immunopath 49:175-186.

Levin S (1987). The immune system and susceptibility to infections in Down's syndrome, in oncology and immunology of Down syndrome. McCoy EE and Epstein CJ, eds., Alan R Liss Inc., New York pp 143-162.

Levin S, Nir E, Mogilner BM (1975). T cell immune-deficiency in Down's syndrome.Pediatrics 56:123-126

Levin S. Schlesinger M, Handzel Z, et al (1979). Thymic deficiency in Down's syndrome. Pediatrics 63:80-86.

Levo Y, Green P (1977). Down's syndrome and autoimmunity. Am J Med Sci 273:95-99.

Lockitch G, Puterman M, Godolphin W et al., (1989). Infection and immunity in Down Syndrome: a trial of long-term low oral doses of zinc. J Pediatr 114:781-787.

Loh RK, Harth SC, Thong YH et al., (1990). Immunoglobulin G subclass deficiency and predisposition to infection in Down's syndrome. Pediatr Infect Dis J 1990; 9:547-551.

Montagna D, Maccario R, Ugazio AG, et al (1988). Cell mediated cytotoxicity in Down syndrome: Impairment of a allogeneic mixed lymphocyte reaction, NK and NK-like activities. Eur J Pediatr 148:53-57.

Oster J, Mikkelsen M, Nielsen A (1975). Mortality and life-table in Down's syndrome. Acta Pediatr Scand 64:322-326

Raziuddin S, Elawad ME (1990). Immunoregulatory CD4+CD45R+ suppressor/inducer T lymphocyte subsets and impaired cell-mediated immunity in patients with Down's syndrome. Clin Exp Immunol 79:67-71.

Richards BW, Sylvester PE (1967). Viral hepatitis and Down's syndrome. The Lancet 1:615.

Schmid F (1983). Cell therapy: A new dimension in medicine. Thun Switzerland: Ott Verlag.

Shibahara Y, Sando I (1989). Congenital anomalies of the eustacian tube in Down's syndrome. Ann Otol Rhinol Laryngol 98:543-547.

Spina CA, Smith D, Korn E et al., (1981). Altered cellular immune functions in patients with Down's syndrome. Am J Dis Child 135: 251-255.

Stark CR, Fraumeni JR., JF (1966). Viral hepatitis and Down's syndrome. The Lancet 1:1036-1037.

Stoller A, Collmann RD (1966). Area relationship between incidences of infectious hepatitis and of the births of children with Down's syndrome nine months later. J Mental Defic Res 10:84-88.

Ugazio AG, Lanzavecchia A, Jayakar S et al., (1978) Immunodeficiency in Down's syndrome. Acta Paediatr Scand 67:705-708.

Van Dyke DC, Lang DJ, Van Duyne S et al., (1990). Cell therapy in children with Down syndrome: A retrospective study. Pediatrics 85:79-84.

# Hematologic and Oncologic Disorders in Down Syndrome

Alvin Zipursky, Annette Poon, and John Doyle

Down syndrome is associated with an increased incidence of leukemia in childhood. This was first noted anecdotally (1) and then by analysis of populations of leukemic children (2). Recent data from Britain (3) indicate that this relationship is still evident (Table 1) and that the incidence of leukemia in Down syndrome in that series is approximately 15 times that found in normal children. That means that approximately one out of every 100 Down syndrome children will develop leukemia.

Initially it was believed that although the incidence was increased, the type of disease was not different. This is erroneous since it has been shown that there is a very high incidence of acute non-lymphoblastic leukemia in Down syndrome (Table 2) and that these cases occur almost exclusively in children less than four years of age (Fig. 1).

It appears now that all of these cases of non-lymphoblastic leukemia in Down syndrome are acute megakaryoblastic leukemia (AMKL), a rare form of leukemia in childhood. A recent study from Japan (5) showed that 14 out of 20 cases of leukemia in Down syndrome were AMKL and occurred in children less than 4 years of age. Similar findings were noted in the Canadian Down Syndrome

Table 1. Incidence of Leukemia in Down Syndrome in Britain
(Levitt et al, Arch. Dis. Child 65:212, 1990)

|  | Found | Expected |
|---|---|---|
| Down Syndrome | 90 | 6.38 |
| Non-Down Syndrome | 4377 |  |
| Total | 4467 |  |

Therefore the incidence of leukemia in Down Syndrome is 14 x greater than in non-Down Syndrome children.

93

**Table 2. Types of Leukemia Found in Down Syndrome (Britain 1986-1990)\***

| Age | 0 | 1 | 2 | 3-4 | 5-9 | 10-14 |
|---|---|---|---|---|---|---|
| ALL | 0 | 0 | 2 | 5 | 8 | 2 |
| ANLL | 4 | 9 | 4 | 1 | 0 | 18 |
| "Unspec." | 4 | 1 | 1 | 0 | 0 | 0 |

"ANLL" in children 3/Total ANLL + ALL = 17/35 = 48%.
*Data from National Registry of Childhood Tumors at the Childhood Cancer Research Group, Oxford, U.K. (4)

Registry (Table 3). These observations, together with those from Britain, indicate that leukemia in Down syndrome is of two types; acute lymphoblastic leukemia (ALL) in older children, and AMKL in children less than 4 years of age. Furthermore, it would appear from all of the above data that approximately 50% of leukemia in Down syndrome is AMKL.

This is remarkable because AMKL is very rare in non-Down syndrome chil-

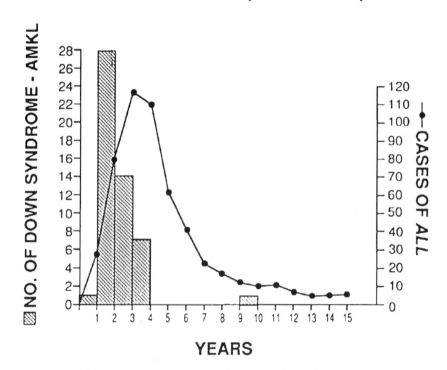

Fig. 1. Age distribution of cases of acute megakaryoblastic leukemia in Down Syndrome. The age distribution of ALL in childhood is shown as the solid line.

Table 3. Canadian Down Syndrome -Leukemia Registry
Acute Lymphoblastic Leukemia in Down Syndrome

|   | Name | Hosp.* | Date of Diagnosis | Age at Diagnosis (months) | FAB | Survival m(months) |
|---|------|--------|-------------------|---------------------------|-----|--------------------|
| 1. | A.R. | M.C.H. | 26/07/90 | 236 | < 1 | — |
| 2. | K.L. | W.C.H. | 06/09/88 | 184 | < 2 | 22+ |
| 3. | K.W. | H.S.C. | 31/07/86 | 34 | < 1 | 53+ |
| 4. | G.M. | H.S.C. | 29/09/80 | 82 | — | 122+ |
| 5. | T.S. | H.S.C. | 13/07/84 | 59 | < 1 | 77+ |
| 6. | Mc.N. | H.S.C. | 29/05/86 | 44 | < 1 | 55+ |

*M.C.H.= Montreal Children's Hospital
W.C.H.= Children's Hospital, Winnipeg
H.S.C.= The Hospital for Sick Children, Toronto
St.J.= Hopital Sainte-Justine, Montreal
C.H.E.O.= Children's Hospital of Eastern Ontario, Ottawa
B.C.C.H.= British Columbia Children's Hospital, Vancouver
A.C.H.= Aberdeen Children's Hospital, Aberdeen, Scotland
N.Y.H.= New York Hospital, Cornell Medical Center
I.W.K.C.H. = Izaac Walton Killam Children's Hospital, Halifax, Nova Scotia
U.S.H.= University Hospital, Saskatoon
— = no data available yet

dren and when it does occur it has an interesting age distribution (Figure 2). One population consists of older children and indeed this form of leukemia is seen in adults. Of interest the other population is in young children with an age distribution similar to that found in Down syndrome.

The clinical features of ALL in Down syndrome are similar to those found in

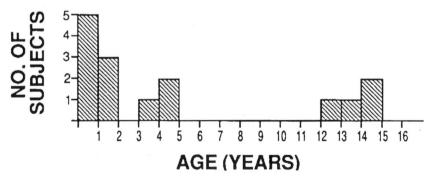

Fig. 2.    Age distribution of cases of acute megakaryoblastic leukemia in non-Down Syndrome children.

other children and it would appear that their response to therapy is also similar. However, in the past it seemed that these patients did very poorly. For example, two years ago, in a conversation with Professor Lejeune of Paris who maintains a large registry of Down syndrome cases in France, he told me that there are no survivors of leukemia in his registry! And yet the cure rate for non-Down syndrome patients is of the order of 70%. This matter was studied in detail recently in Britain (3), where it was also found that the cure rate was significantly lower than normal (20% vs. 50%). That study showed that the low cure rate was partly a result of either inadequate treatment or no treatment at all.

Recent observations from the Canadian Down Syndrome-Leukemia Registry indicate that many of these patients can be long-term survivors with appropriate therapy (Table 4). The results of treatment of Down syndrome patients with ALL within the Children's Cancer Study Group of the United States also show that the response to therapy in these children is not different than that of other children (2).

The clinical features of AMKL have been reviewed previously by us (6) and consist of a pre-leukemic phase of about six months, during which the disease appears insidiously, often as thrombocytopenia and then gradually evolves into overt leukemia with large numbers of primitive cells in the blood and bone marrow. There is often an associated fibrosis of the bone marrow.

The leukemic cells in this disease have been identified as megakaryoblasts by virtue of their morphology on light microscopy, ultrastructure and immunologic markers (5,7). It is of interest that these megakaryoblasts also have features of early erythroid precursors (8). Indeed it has been reported that Down syndrome children have an abnormally high incidence of erythroleukemia (9) and con-

Table 4. Canadian Down Syndrome -Leukemia Registry
Acute Non-Lymphoblastic Leukemia in Down Syndrome

|   | Name | Hosp.* | Date of Diagnosis | Age at Diagnosis (months) | FAB | Resp. to Rx. | Survival |
|---|------|--------|-------------------|---------------------------|-----|--------------|----------|
| 1. | R.H. | H.S.C. | 15/05/86 | 26 | M7 | CR | 54+ |
| 2. | B.B. | St.J. | 24/04/89 | 39 | M7 | CR | 13+ |
| 3. | T.D. | C.H.E.O. | 14/08/90 | 21 | M6 | CR | 4+ |
| 4. | I.S. | B.C.C.H. | 11/01/90 | 14 | M6-7 | — | — |
| 5. | K.S. | A.C.H. | 31/08/90 | 28 | M6-7 | NR | 1 (died) |
| 6. | N.M. | I.W.K.C.H. | 25/08/90 | 21 | M6-7 | — | — |
| 7. | C.W. | N.Y.H. | 26/09/90 | | M7 | — | — |
| 8. | K.M. | W.C.H. | 17/03/86 | 25 | "AML" | CR | 52+ |
| 9. | D.C. | W.C.H. | 30/05/84 | 13 | M7 | No Rx | — |
| 10. | E.P. | U.S.H. | 22/05/90 | 19 | M7 | — | — |

*as for Table 3

versely in many of these cases the leukemic erythroid cells have properties of megakaryoblasts (10). We conclude therefore, that this form of leukemia in Down syndrome represents a proliferation of a biphenotypic hematopoietic progenitor cell with potential to form both megakaryocytes and erythroid precursors. Such cells have been isolated from normal bone marrow (11). The nature of these malignant cells and their relation to a benign proliferation of megakaryoblasts which also is seen in Down syndrome children (see below) are the topic of much study not only by my colleagues and myself but also by other groups in the world. AMKL in Down syndrome is curable as shown in Table 3, which displays the results of the Canadian Down Syndrome Registry. Also the studies of the Children's Cancer Study Group of the United States indicate that the results of treatment may actually be better in Down syndrome than in other children (6).

Thus in children with Down syndrome there is an abnormally high incidence of a malignant disease characterized by a proliferation of megakaryoblasts. Of great interest in this regard is that in newborn infants with Down syndrome there occurs a remarkable disorder, transient leukemia (TL), in which there is a benign proliferation of megakaryoblasts.

We have reviewed this disorder previously (6); it is found exclusively in newborn infants with Down syndrome and usually has been found in a well child who happens to have had a blood count done for other reasons whence it is found that there is an abnormal number of primitive cells in the blood. These cells have the properties of megakaryoblasts as judged by their ultrastructural and im-munologic features (12,13). There are reports indicating that the biological prop-erties of these cells differ from those of the leukemic cells in AMKL (14), however, there is need for further study to characterize and compare these benign and malignant proliferations.

In our previous review (6) we suggested that some of these children with transient leukemia (a disorder that disappears in the first few months of life) may go on to develop AMKL in the first four years of life. We estimated that 20% of patients with transient leukemia may eventually develop AMKL. Recently Dr. Alan Homans of Brown University carried out a questionnaire survey within the two major Children's Cancer Study Groups in the United States, CCSG and POG. He obtained information on 62 cases of transient leukemia and found that 30% had developed leukemia in the first 3 years of life (15). Thus, the benign proliferation of megakaryocytes seen in transient leukemia also in some cases reprerents a pre-leukemic disorder.

Our experience with transient leukemia is shown in Table 5 which contains information about cases included in the Canadian Down Syndrome-Leukemia Registry. In all cases the megakaryoblasts disappeared from the blood within the first two months of life. There is another interesting feature of this disorder. This was reported recently by Iselius et al from Southampton (16). They found that the age of mothers of infants with transient leukemia was significantly lower than mothers of infants with Down syndrome and leukemia. At first glance this

**Table 5. Canadian Down Syndrome -Leukemia Registry
Transient Leukemia in Down Syndrome**

| Patient | Hosp.* | D.O.B. | WBC (mm3) | Blasts (mm3) | platelets (mm3) |
|---------|--------|--------|-----------|--------------|-----------------|
| G.H. | St.J. | 26/04/90 | 59,700 | — | 68,000 |
| K.A.D. | B.C.C.H. | 05/10/90 | 14,600 | 438 | 65,000 |
| S.T.J. | W.C.H. | 18/04/90 | 14,000 | 3360 | 19,000 |
| M. | H.S.C. | 13/08/86 | 47,670 | 22,900 | 104,000 |
| L.P. | H.S.C. | 23/11/86 | 17,900 | 1,970 | 48,000 |
| C. | H.S.C. | 09/05/87 | 52,900 | 21,840 | 270,000 |
| G. | H.S.C. | 15/07/87 | 8,000 | 2,300 | 105,000 |
| H. | H.S.C. | 02/10/88 | 70,100 | 40,660 | 308,000 |
| R. | H.S.C. | 04/02/89 | 31,800 | 8,820 | 50,000 |
| L. | H.S.C. | 03/07/89 | 14,500 | 1,440 | 315,000 |
| B.M. | H.S.C. | 20/07/89 | 108,000 | 60,550 | 441,000 |
| Z. | H.S.C. | 26/02/90 | 70,980 | 57,490 | 147,000 |
| P. | H.S.C. | 07/08/90 | 31,870 | 9,880 | 441,000 |
| T. | H.S.C. | 04/09/90 | 11,290 | 900 | 21,000 |

*as for Table 3

suggests that Down syndrome children who develop transient leukemia are drawn from a different group than those who do not. If correct, this would be a remarkable observation. There are however other possible interpretations of these data and their conclusions must await further studies.

We have studied the blasts in these cases and they have the properties of megakaryoblasts. These properties include morphology, immunocyto-chemistry and ultrastructure (6). Although many of the blasts have characteristics of the primitive blasts seen in AMKL ,some show differentiation towards a megakaryocyte (12-14).

There are interesting hematologic abnormalities associated with transient leukemia. Not unlike the situation in AMKL, there are abnormalities in the erythroid series. We have observed morphologic features of dyserythropoiesis in these cases and this has also been associated with staining of red cell precursors with periodic acid-Schiff stain which is found in dyserythropoiesis (17). Also it has been reported that the blasts in these patients produce basophils in culture, and some of the megakaryoblasts in vivo contain basophil granules (18). The association of these three lines of blood cells, megakaryoblasts, erythroid precursors, and basophils is of interest. As noted above in AMKL there is a biphenotypic precursor cell which has the potential of forming either megakaryocytes or erythroid cells. It is possible therefore that the abnormal cell that proliferates in transient leukemia is a relatively undifferentiated precursor cell that has trilineage potential. We have seen one patient with transient leukemia who at three weeks of age, while recovering from transient leukemia, developed basophilia. One similar observation has been recorded previously (13). Parenthetically, it has been noted that in AMKL the blast cells in culture produce basophils (19). We have recently confirmed that observation (20). It has been reported also that there is a DNA-binding protein that stimulates production of erythroid, megakaryocytic and mast cell elements (21).

We have recently seen a case of transient leukemia in a stillborn fetus with Hydrops Fetalis. The white blood cell count was grossly elevated and the baby had hepatosplenomegaly. There have been several other reports of Hydrops in transient leukemia. Two cases were reported (22) in which a hydropic Down syndrome child was born with an elevated leukocyte count and on postmortem examination was found to have extensive fibrosis throughout the body. We also have seen two such cases; the one referred to above and another that was seen several years earlier.

There has also been a report from Germany of two fetuses who died of hydrops and were found to have trisomy 21 (23). That report was interesting also because it suggested that the incidence of transient leukemia may be much higher than has been suggested previously since in that study 2 out of 24 fetuses, diagnosed by cord as Down syndrome, were found to have transient leukemia.

We have conducted a study recently to determine the frequency of transient leukemia of newborn infants with Down syndrome. In several hospitals in Toronto all newborn infants with Down syndrome are screened for the presence of megakaryoblasts in their blood. To date we have studied 25 infants and in five we have found megakaryoblasts in their blood.

The frequent finding of megakaryoblasts in the blood of newborn infants suggests that this disorder of hematopoiesis is a relatively common phenomenon. There is other evidence that hematopoiesis is abnormal in Down syndrome. Thus, it has been reported that polycythemia is more common in newborns with Down syndrome than in normals (24). Also, the average hematocrit at birth is significantly higher in Down syndrome infants than in normals (25).

Platelets in Down syndrome have been the subject of considerable study and it is clear that they are different than the platelets of normal children. In addition, it has been reported that platelet numbers are lower in the Down syndrome newborns than in the normals (26).

There is evidence of abnormalities in the myeloid system in Down syndrome. It has been reported that the number of granulocyte progenitor cells in the blood of patients with Down syndrome is abnormally low (27). Also, there have been numerous reports of abnormalities in circulating granulocytes including morphology (28), enzyme levels (29), and ability to kill bacteria (30). It is not clear if these phenomena are related to any of the other hematologic abnormalities described above, nor it is known that this influences the susceptibility to infection that has been noted in Down syndrome children.

There are also numerous reports of abnormalities in the immune system in Down syndrome including both cellular and humoral immunity (31). This will not be reviewed here but has been extensively studied and reviewed elsewhere (31).

Finally there is evidence that in the siblings of Down syndrome there is an increased incidence of leukemia and other cancers (32). In that same study it was noted that the incidence of Down syndrome was abnormally high in the siblings of non-Down syndrome children who had leukemia. Thus, there appears to be a genetic relationship between the development of leukemia and of Down syndrome.

The biology of these conditions is, of course, of great interest and we as well as others are studying it. We believe that there is an opportunity to study the relation between a benign and a malignant proliferation of megakaryoblasts and how the benign proliferation progresses into a malignant process. We hope to determine the mechanism whereby a specific genetic defect, namely trisomy 21, can lead to these hematologic disorders. The answer to that question should shed light on the nature of leukemia and the mechanism whereby a third chromosome number 21 leads to the congenital defects associated with Down syndrome.

## REFERENCES

1. Krivit W and Good RA. Simultaneous occurrence of mongolism and leukemia. Amer J Dis Child 94:289, 1957.
2. Robison LL, Nesbit ME, Harland, N et al. Down syndrome and acute leukemia in children: a 10-year retrospective survey from Children's Cancer Study Group. J Pediat 105:235, 1984.
3. Levitt GA, Stiller CA and Chessells JM. Prognosis of Down's syndrome with acute leukemia. Arch Dis Child 65:212, 1990.
4. Stiller CA. Personal communication.
5. Kojima S, Matsuyama T, Sato T, et al. Down's syndrome and acute leukemia in children: an analysis of phenotype by use of monoclonal antibodies and electron microscopic platelet peroxidase reaction. Blood 76:2348, 1990.
6. Zipursky A, Peeters M and Poon A. Megakaryoblastic leukemia and Down's syndrome. Pediat Hemat Oncol 4:211, 1987.
7. Bevan D, Rose M, and Greaves M. Leukemia of platelet precursors: diverse features in four cases. Brit J Haemat 51:147, 1982.
8. Breton-Gorius J, Villeval JL, Kieffer N, et al. Limits of phenotypic markers for the diagnosis of megakaryoblastic leukemia. Blood Cells 15:299, 1989.
9. Malkin D and Freedman M. Childhood erythroleukemia: Review of clinical and biological features. Amer J Pediat Hemat/Oncol 11:348, 1989.
10. Debili N, Kieffer N, Mitjavila MT, et al. Expression of platelet glycoprotein by erythroid blasts in four cases of trisomy 21. Leukemia 3:669, 1989.
11. Nishi T, Nakahata T, Koike K, et al. Induction of mixed erythroid-megakaryocyte colonies and bipotential blast cell colonies by recombinant human erythropoietin in serum-free culture. Blood 76:1330, 1990.
12. Coulombel L, Derycke M, Villeval JL, et al. Characterization of the blast population in two neonates with Down's syndrome and transient myeloproliferative disorder. Brit J Haemat 66:69, 1987.
13. Suda J, Eguchi M, Ozawa T, et al. Platelet peroxidase-positive blast cells in transient myeloproliferative disorder with Down's syndrome. Brit J Haemat 68:181, 1987.

14. Eguchi M, Sakakibara H, Suda J, et al. Ultrastructural and ultracytochemical differences between transient myeloproliferative disorder and megakaryoblastic leukemia in Down's syndrome. Brit J Haemat 73:315, 1989.
15. Homans A. Personal communication.
16. Iselius L, Jacobs P and Morton N. Leukaemia and transient leukaemia in Down syndrome. Hum Genet 85:477, 1990.
17. Quaglino D and Hayhoe FGJ. Periodic acid-Schiff positivity in erythroblasts with special reference to Di Guglielmo's syndrome. Brit J Haemat 6:26, 1960.
18. Suda J, Eguchi M, Ozawa T, et al. Platelet peroxidase-positive blasts in transient myeloproliferative disorder with Down's syndrome. Brit J Haemat 68:181, 1988.
19. Suda T, Suda J, Miura Y, et al. Clonal analysis of basophil differentiation in bone marrow cultures from a Down's syndrome patient with megakaryoblastic leukemia. Blood 66:1278, 1985.
20. Doyle JD, Poon A and Zipursky A. Unpublished observations.
21. Martin DIK, Zon LI, Mutter G, et al. Expression of an erythroid transcription factor in megakaryocytic and mast cell lineages. Nature 344:444, 1990.
22. Becroft DMO and Zwi J. Perinatal visceral fibrosis accompanying the megakaryoblastic leukemoid reaction of Down syndrome. Pediat Path 10:397, 1990.
23. Zerres K, Schwanitz G, Niesen M, et al. Prenatal diagnosis of acute non-lymphoblastic leukemia in Down syndrome. Lancet 335:117, 1990.
24. Weinberger MW and Oleinick A. Congenital marrow dysfunction in Down's syndrome. J Pediat 77:273, 1970.
25. Miller M and Cosgriff JM. Hematological abnormalities in newborn infants with Down syndrome. Am J Hum Genet 16:173, 1983.
26. Thüring W and Tönz O. Neonatale thrombozytenwerte bei kindern mit Down-syndrome and anderen autosomalen trisomien. Helv Paediat Acta 34:545, 1979.
27. Standen G, Philip MA and Fletcher J. Reduced number of peripheral blood granulocytic progenitor cells in patients with Down syndrome. Brit J Haemat 42:417, 1979.
28. Mittwoch V. Frequency of drumsticks in normal women and in patients with chromosomal abnormalities. Nature 201:317, 1964.
29. Mellman WJ, Oski FA, Tedesco TA, et al. Leucocyte enzymes in Down's syndrome. Lancet 2:674, 1964.
30. Kretschmer RR, Lopez-Osuna M, De La Rosa L, et al. Leukocyte function in Down's syndrome: quantitative NBT reduction and bacteriacidal capacity. Clin Immunol and Immunopath 2:449, 1974.
31. Ganick DJ. Hematological changes in Down's syndrome. CRC Crit Rev Onc/Hem 6:55, 1986.
32. Miller RW. Down's syndrome (Mongolism), other congenital malformations and cancers among the sibs of leukemic children. New Eng J Med 268:393, 1963.

# Neurological and Neurobehavioral Disorders in Down Syndrome

Ira T. Lott, MD

The central nervous system is the most commonly affected organ system in Down syndrome. Although there is still not a complete understanding of the mechanism of the ubiquitous developmental disability in this disorder, there have been advances in the understanding of brain function in individuals with Down syndrome.

## STRUCTURAL STUDIES OF BRAIN

In his comprehensive review of the neurology of Down syndrome in 1977, Zellweger summarized the gross neuropathological changes as comprising slightly low brain weight, shortened occipital-frontal diameter, steep occipital ascent, and hypoplasia of the operculum and superior temporal gyrus. At the time of previous reviews (Lott, 1982, 1986), very little information was available in regard to a systemic microscopic description of the disorder. Emerging data suggests a defect in cerebral granular layers II and IV, basal ganglia calcification, and a myelination delay (Wisniewski, 1990). Abnormalities in synaptic density appear to occur late in gestation and probably continue after birth (Takashima, 1981). These changes may be responsible for a deceleration in brain growth after infancy accounting for the microcephaly seen later in life. In our experience, microcephaly is usually mild in children and adults with Down syndrome. Severe departures from the normal range generally connote another pathological process additive to trisomy 21.

Whether these findings are specific to Down syndrome cannot be stated with certainty at present. For example, Hughes et al. (1991) reviewed 2320 neonatal cranial sonograms and found evidence of heightened thalamic echogenicity, a possible precursor to basal ganglionic calcification, in a variety of congenital and acquired disorders including Down syndrome. Huttenlocher (1984) has described alterations in synaptic density and morphology in other developmental disabilities than Down syndrome. Nonetheless, the evidence to date suggests a defect in neuronal migration as well as a problem in postnatal synaptogenesis as morphological determinants of the defect in cognitive function seen in children with Down syndrome. As reviewed by McCoy and Enns (1986), either primary or secondary disturbances have been seen in dopaminergic, serotonergic, and cholinergic neurotransmitter systems in Down syndrome. Florez et al., (1990) reported a marked receptor reduction in muscarinic receptors in the midbrain of 2 stillborn infants with trisomy 21 even though conventional histology was normal. Perhaps the most dramatic structural abnormality in Down syndrome at the level of the neurological phenotype is the occurrence of Alzheimer disease neuropathology in the brains of individuals over age 35 years. We have accorded a special discussion for this entity (see chapter by Wisniewski).

## FUNCTIONAL STUDIES IN BRAIN

Penrose observed in his 1960 monograph that mental retardation was the most constant feature of Down syndrome. Moore (1973) found that only 2 of 2750 individuals with Down syndrome had an IQ measure greater than 85 with 7 scoring in the borderline range from 70-84. The cognitive dysfunction in Down syndrome is not homogenous but shows a particular deficit in the areas of auditory and visual sequential memory (Marcell and Armstrong, 1982) as well as systematic difficulties in speech and language skill development (Fowler, 1990 and see chapter by Miller). It is still not clear as to whether the progressive deceleration in brain growth reflected in the microcephaly discussed above is associated with an invariable decline in cognitive and adaptive functioning. Silverstein et al., (1988) suggest that motor competence may decline in individuals with Down syndrome overage 60 years but that other regression is not seen invariably. This suggests that the dementia of Alzheimer disease in Down syndrome is the reflection of a specific disease process in some individuals and not universally part of the aging phenomena.

How the disorders in brain function in Down syndrome relate to the trisomic portion of the 21st chromosome has been the subject of intense discussion. (Epstein, Korenberg, Lott et al.,1991). Phenotypic studies of partial aneuploidy support the Epstein (1990) hypothesis that a triple dose of the critical region of chromosome 21 causes anomalies of organogenesis with gene products sufficient to account for the clinical changes. This would hold for brain as well as for other organ systems. In a slightly different interpretation, Opitz and Gilbert-Barness (1990) suggest that the trisomic state induces a perturbation in the factors controlling the rate and timing of developmental sequences and this is the key to understanding the phenotype. The latter theory is supported by the observations that there is not a single phenotypic anomaly within any organ system in Down syndrome which is specific to the disorder but that the characteristic clinical state reflects a disorder of sequencing and timing of development. The dysmaturity in the temporal lobe may correlate with the memory difficulties described above.

## SEIZURE DISORDERS

The reports of seizures in individuals with Down syndrome have shown an incidence ranging from 1% (Kirman, 1951) to 13.6% (Paulson et al., 1969). The number of patients in a given study has varied widely and the concurrence of Down syndrome with other neurological disorders responsible for the production of seizures has not always been clear. Romano et al., (1990) found an overall incidence of 13.2% in a group of 113 individuals with Down syndrome formed from institutional and community environments. Febrile seizures were equal in incidence to the general population but there was a relative increase in the incidence of afebrile seizures and infantile spasms. Pueschel et al., (1991) have reported a bimodal incidence of seizures in Down syndrome with peaks before 1 year of age and the second peak after the third decade. This group also noted a relative increase in infantile spasms in infants with Down syndrome. Lott and Lai, (1981) have noted an increased incidence of seizures early in the course of the dementia of Alzheimer's disease in individuals with Down syndrome. In general, seizures do not occur early in the course of Alzheimer's disease in the general population. Seizures alone in older individuals with Down syndrome may impair attention and cause a pseudo-dementia with loss of adaptive capacities. Happily, this problem is often corrected without difficulty by anticonvulsant therapy. In general, seizure control in Down syndrome is not complicated unless there is a second neurological insult responsible for additional brain damage.

## HYPOTONIA

Hypotonia in the infant with Down syndrome was noted in the 1920s and has been part of the clinical presentation of this condition in subsequent reports. The hypotonia appears to have a central origin (Brousseau and Brainerd, 1928) and may reflect an immaturity of integrating kinesthetic, tactile, and cutaneous inputs. Others have cited an immaturity of oculomotor programming. There appears to be an inverse relationship between the degree of hypotonia and performance on cognitive (Cichetti and Sroufe, 1976) and adaptive (Cullen et al., 1981) tasks. The hypotonia remits slowly throughout childhood but there may be some increase in muscle activation though passive tactile stimulation (Linkous and Stutts, 1990). Hoyer and Limbrock (1990) have suggested that early functional training of the orofacial muscles may counteract the hypotonia affecting open mouth expression, tongue prolapse, and everted lower lip seen in some infants with Down syndrome.

## ATLANTOAXIAL INSTABILITY

The hypotonia described above leads to a general ligamentous laxity which, for the Down syndrome child, carries most ominous concerns for the atlantoaxial junction. Since the incidence of atlantoaxial instability in Down syndrome ranges from 10-20%, as many as 100,000 individuals with the disorder may have this condition (Cooke, 1984). The most important structure involved appears to be the transverse atlantal ligament which forms the posterior support of the odontoid process. Actual dislocations of the atlantoaxial junction may compress the cervical spinal cord with symptoms of neck pain, head tilt, torticollis, abnormal gait, and paraplegia. The American Academy of Pediatrics has recommended (1984) that all children with Down syndrome who wish to participate in sports that involve possible trauma to the head have lateral x-ray films of the neck in neutral, flexion, and extension. If the distance between the odontoid process of the axis and the anterior arch of the atlas exceeds 4.5mm or the odontoid is abnormal, participation in sports should be restricted. The orthopedic issues regarding this entity are discussed elsewhere in this symposium (see chapter by Diamond). Davidson has recently evaluated the data accumulating in regard to neurological consequences of atlantoaxial instability in Down syndrome (1988) and raises the question to what extent instability on x-rays may or may not indicate the likelihood of developing dislocation. He argues that the great majority of individuals with Down syndrome who develop a neurologically significant dislocation had signs and symptoms occurring several months before the actual event. At any rate, longitudinal studies of this entity are now needed with particular emphasis on the female preponderance insusceptibility.

## NEUROBEHAVIORAL ASPECTS OF DOWN SYNDROME

Since the early descriptions of this entity (Down, 1887), the personality characteristics of individuals with Down syndrome have been stated to include a power of imitation, a strong sense of the ridiculous, and a great obstinacy for individual actions. Silverstein et al., (1985) employed checklists for evaluating personality characteristics of institutionalized and community based individuals with Down syndrome and compared them to appropriate control groups. They confirmed the view that, as a rule, children with Down syndrome appear to be outgoing, affectionate, with social competence often out of proportion to cognitive functioning. As recently reviewed by Sigler and Hodapp (1991), the social quotients of these children usually exceed their IQs between the ages of 4-17 years, but as higher order linguistic skills are required, social and adaptive behaviors level off (Cornwell and Birch, 1969).

Lund (1988) studied 324 individuals with mental retardation of whom 44 were diagnosed as Down syndrome and found that while women with Down syndrome had few psychiatric symptoms, men with Down syndrome had an increased prevalence of psychiatric, behavioral, neurotic and deviant social problems. The study gave no idea as to how to explain these differences. None of the Down syndrome group studied showed schizophrenic or affective disorders, although individual cases with these problems have been occasionally described. In reviewing psychiatric disorders in adolescents and young adults with Down syndrome, Harris (1988) stresses the risk these individuals have for personal losses in this age group. Indeed, reactive depression may be mistaken for the early dementia of Alzheimer disease in the young adult with Down syndrome.

The association of Down syndrome with other neuropsychiatric disorders is not completely understood. In part, the frequency of Down syndrome is sufficiently high that a fortuitous occurrence with other entities is possible. The author has personally examined an individual with Down syndrome, Klinefelter's syndrome, and tuberous sclerosis. On the other hand, there does appear to be an increased incidence of infantile autism and Down syndrome with up to 1 in 20 children with Down syndrome having autistic features (Gillberg et al, 1986). Perhaps there is an increased brain susceptibility in Down syndrome to acquired disorders that result inautistic symptomatology. Attwood et al. (1988) argue that it is the intentional expression of feeling states through gestures that qualitatively differentiates Down syndrome from autistic adolescents and that the deficit inautism is syndrome specific. Ismail (1988) has reported Tourettism in an individual with Down syndrome adding this report to two other cases.

## CONCLUSIONS

This chapter reviews neurological aspects of Down syndrome with the exception of Alzheimer's disease, which is discussed separately in this volume. Structural studies in brain are consistent with a defect in neuronal migration and a disorder of postnatal synaptogenesis. Mental retardation is ubiquitous within the disorder with a particular disability in the area of auditory and verbal sequential memory. Seizure disorders in Down syndrome occur in up to 13% of individuals with a bimodal incidence and increase in the frequency of infantile spasms. Hypotonia in infants and children with Down syndrome appears to be central in origin and the resultant ligamentous laxity accounts for the high degree of atlantoaxial instability. In the early years, social quotients advance faster than IQ in children with Down syndrome. In later years, men have a higher frequency than women of behavioral and neuropsychiatric disorders. There appears to be a higher than chance association between Down syndrome and infantile autism.

## REFERENCES

American Academy of Pediatrics, Committee on Sports Medicine: (1984). Atlantoaxial instability in Down syndrome. Pediatrics 74:152-154.

Attword, Anthony, Frith U, Hermelin (1988). The Understanding and Use of Interpersonal Gestures by Autistic and Down's Syndrome Children. Jour of Autism and Developmental Disorders 18:241-257.

Cichetti D and Sroufe LA (1976). The relationship between affective and cognitive development in Down's Syndrome Children. Child Dev 47:920-929.

Collacott A and Ismail IA (1988). Tourettism in a patient with Down's Syndrome, Jour of Ment Deficiency 32:163-166.

Cooke RE (1984). Atlantoaxial instability in individuals with Down's syndrome. Adapt Phys Activ Qu 1:194-196.

Cornwell A and Birch H (1969). Psychological and social development in home-reared children with Down's syndrome (mongolism), Amer Jour Ment Deficiency 74:341-350.

Cullen S, Cronk C, Pueschel S, Schnell R, Reed R. (1981). Social Development and Feeding Milestones of Young Down Syndrome Children. Am Jour Ment Defic 85:410-415.

Davidson R (1988). Atlantoaxial Instability in Individuals with Down Syndrome: A Fresh Look at the Evidence. Pediatr 81:857-865.Down, JLH (1987). "Mental Affections of Childhood and Youth". London: Churchill.

Epstein, Charles J (1990). The Consequences of Chromosome Imbalance. Amer Jour of Med Genetics Suppl 7:31-37.

Florez J, del Arco C, Gonzalez A, Pascual J, Pazos A (1990). Autoradiographic Studies of Neurotransmitter Receptors in the Brain of Newborn Infants with Down Syndrome. Amer Jour of Med Gen Suppl 7:301-305.

Fowler A. (1990). Language abilities of children with Down syndrome: evidence for a specific syntactic delay. In: Cicchetti D and Beweghly M (eds), "Children with Down Syndrome: A Developmental Perspective", Cambridge: Cambridge University Press.

Gilberg C, Persson E, Grufman M, Themner U. (1986). Psychiatric disorders in mildly and severely mentally retarded urban children and adolescents: epidemiological aspects. Br J Psychiatry 149:68-74.

Harris James (1988). Psychological Adaption and Psychiatric Disorders in Adolescents and Young Adults with Down Syndrome. In Pueschel SM (ed.) "The Young Person with Down Syndrome", Baltimore, MD: Paul H. Brooks Publishing Co.

Hoyer H and Limbrock GJ (1990). Orofacial regulation therapy in young children with Down syndrome, using methods and appliances of Castillo-Morales. Asdc Jour of Dentistry 57:442-444.

Hughes P, Weinberger E, Shaw DW (1991). Linear areas of echogenicity in the thalami and basal ganglia of neonates an expanded association. Radiol 179(1):103-105.

Huttenlocher P (1984). Synapse elimination and plasticity in developing human cerebral cortex. Am Jour Ment Defic 88:488-496.

Kirman BH (1951). Epilepsy in mongolism. Arch Dis Child 26:501-503.Linkous LW and Stutts RM (1990). Passive tactile stimulation effects on the muscle tone of hypotonic, developmentally delayed young children. Perceptual and Motor Skills, 71:951- 954.

Lott IT and Lai F (1982). Dementia in Down Syndrome—observations in a Neurology Clinic. Applied Res Ment Retard 3:232-239.

Lott IT, (1986). Neurology of Down Syndrome. In: Epstein CJ (ed), "The Neurobiology of Down Syndrome". New York, Raven Press, pp. 17-28.

Lund J (1988). Psychiatric aspects of Down's syndrome, Acta Psychiatr Scand 78:369-374.

McCoy EE, Enns L (1986). Current status of neurotransmitter abnormalities in Down Syndrome. In: Epstein CJ (ed), "The Neurobiology of Down Syndrome", New York: Raven Press, pp. 73-87.

Marcell MM and Armstrong V (1982). Am Jour of Ment Deficiency 87:86-95.Moore BC, (1973). Am Jour of Ment Deficiency 17:46-51.

Opitz JM and Gilbert-Barness EF (1990). Reflections on the Pathogenesis of Down Syndrome, Amer Jour of Med Genetics Suppl 7:38-51.

Paulson GW, Son CD, Nance WE (1969). Neurological aspects of typical and atypical Down's syndrome. Dis Nerv Syst 30:632-636.Penrose LS (1963). "The Biology of Mental Deficiency" 3rd ed, New York: Grune and Stratton.

Pueschel Sm, Louis S, McKnight P (1991). Seizure disorders in Down Syndrome. Archives of Neurology 48:318-320.

Romano C, Tine A, Gazio G, Rizzo, R, Colognola R, Sorge G, et al. (1990). Seizures in Patients with Trisomy 21. Amer Jour of Med Gen 7:298-300.

Silverstein AB, Ageno D, Alleman AC, Derecho KT and Gray S (1985). Amer Jour of Ment Deficiency 89:555-558.

Silverstein AB, Herbs D, Miller TJ, Nasuta R, Williams DL, White JF. (1988). Effects of age on the adaptive behavior of institutionalized and noninstitutionalized individuals with Down syndrome. Amer Jour of Ment Retard 92(5):455-460.

Takashima S, Becker LE, Armstrong DL, Chan F (1981). Abnormal neuronal development in visual cortex of the human fetus and infant with Down's syndrome: A quantitative and qualitative Golgi Study. Brain Res 225:1-21.

Wisniewski KE (1990). Down Syndrome Children Often Have Brain with Maturation Delay, Retardation of Growth, and Cortical Dysgenesis. Amer Jour of Med Genetics Suppl 7:274-281.

Zellweger H. (1977). In: Vinken PJ and Bruyn GW (eds.), "Handbook of Clinical Neurology, Congenital Malformation of the Brain and Skull", Amsterdam, Holland, Vol. 31, pp. 367-470.

Zigler E and Hodapp R (1990). Behavioral Functioning in Individuals with Mental Retardation. Annu Rev Psychol 42:29-50.

# Orthopedic Disorders in Down Syndrome

Liebe S. Diamond, MD

When patients with Down Syndrome succumbed early in life to congenital heart disease, infections, and tumors, there was very little emphasis upon or definition of their disabling skeletal problems. As a consequence of the increase in life expectancy, orthopedic surgeons have been compelled to explore the natural history of skeletal disease in Down Syndrome and to develop a rational system for rehabilitation. These patients have poor connective tissue, thin fascial layers, tenuous muscle and tendon insertions, and poor capsular tissues about major joints. Rehabilitative surgery must be special in design and meticulous in execution, while post-operative management and physical therapy must be prolonged and intense. The surgeon must be mindful of the strange and sometimes surgically defeating postures in which patients with Down Syndrome sit and stand. (Fig.1).

Because of the underlying quality of tissues in these patients, the success rate for any surgical procedure is diminished by a significant percentage when compared with the general population. It is therefore wise for the orthopedic surgeon to explain all of the alternatives very carefully and to prepare a detailed certificate of informed consent for each individual patient, spelling out the problem, the goals of the surgical procedure, and the relative risk of being unable to achieve that goal. The care giver—be it parent or institutional staff—must be aware that the rehabilitation process will be prolonged.

At the same time, one must state that the vast majority of patients with Down Syndrome can be helped and will do reasonably well after surgery.

No patient who has a disabling orthopedic problem should be turned away simply because he has Down Syndrome. These patients have wonderful personalities and respond to gentle, persistent, consistent, and personally attentive post-operative physical therapy. It is sometimes necessary to restrain overly enthusiastic parents and therapists from pushing the patient forward too rapidly. Because of the slow rate of healing of fibrous and capsular tissues, it is best to double the time for each stage of post-surgical rehabilitation.

The orthopedic problems encountered in Down Syndrome occur most commonly in the foot, knee, hip and spine. These problems will be discussed in descending order of their frequency.

## FEET

The first visit to the orthopedic surgeon usually comes at age 2 with the chief complaint that the child is not yet walking. Parents express concern about position of the feet and legs. An infant stimulation program with developmental physical therapists can move things along a bit. However, these babies are so floppy that it take them a long time to get everything in balance. An 18 month old child with Down Syndrome is rather like a 3 week old puppy with all four limbs going in different directions, loose, floppy, and having a wonderful time.

At about 2½ to 3 years of age most of these children will be able to walk reasonably well. It is at this time that we see the emergence of a characteristic foot deformity which includes a fat, flat foot with a "hitchhiking" big toe and soft, doughy, very dry skin. At this time we also see the first evidence of metatarsus primus varus and hallux valgus. By adolescence, this deformity can increase significantly. Growing body weight produces splaying of the foot and pressure of the shoe converts the metatarsus primus varus and hallux valgus into the tough combination of metatarsus primus varus and hallux valgus. For this reason it is important for patients with Down Syndrome to wear shoes which are of adequate width to prevent excessive pressure on the great toe. A reasonably protective shoe prevents the occurrence of chilblains which result from the chronically cold, wet feet (Fig. 2,3,4) associated with canvas sneakers and winter weather.

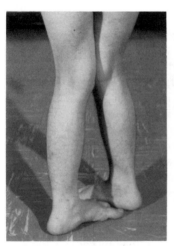 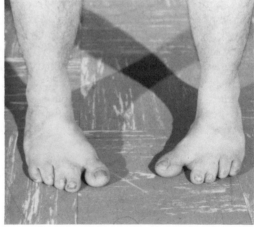

Fig. 1. Bizarre stance with hallux varus and dislocated patellae.     Fig. 2. Metatarsus primus varus and hallux varus.

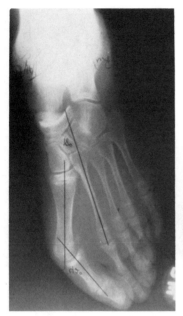 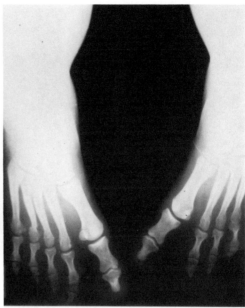

Fig. 3. Metatarsus primus varus Fig. 4. Metatarsus varus and hallux varus.
with hallux valgus (bunion).

The pronated feet of patients with Down Syndrome are not a major problem. Only those patients with foot or leg pain require treatment for their flat feet. There is no cure for the deformity except surgically and this should be reserved for those patients who are symptomatic and unrelieved by arch supports and carefully chosen shoes. Patients with hallux valgus have accentuation of the deformity when they walk in severe pronation and require flexible arch supports as part of their pre- and post-operative treatment. Orthopedic shoes with steel shanks are never indicated in these patients. The weak muscles and plantar structures make it impossible for patients with Down Syndrome to flex the sole of the rigid shoe, leading to sore feet. It is best to select a flexible shoe and add scaphoid pads and a medial heel wedge. If the patient is comfortable and does not have callouses or sore spots and the toes are not cramped, the shoe is appropriate.

For those patients with severe metatarsus primus varus and hallux valgus, early surgical correction of the deformity will prevent secondary severe hallux valgus. (Fig. 5) Soft tissue procedures alone are of no use in the feet in Down Syndrome. Basilar osteotomies of the first metatarsal heal well and provide an excellent and serviceable foot. Distal osteotomies of the first metatarsal do not provide the necessary angular correction, are slow to heal and (Fig. 6) lead to an S-shaped first metatarsal which is difficult to fit in a shoe. Most patients who come to surgery for their forefoot deformity do so in late childhood or early adolescence. This does not preclude offering treatment to symptomatic older patients. Surgery includes

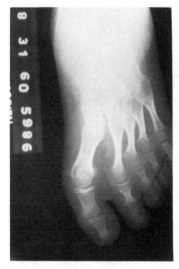

Fig. 5. Early bunion.

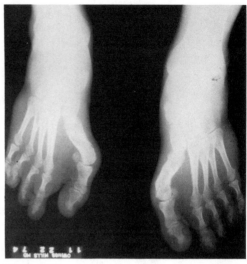

Fig. 6. S-shaped 1st metacarpal after failure of distal osteotomy for bunion.

correction of the hallux valgus by capsuloplasty and adductor hallucis release coupled with first metatarsal osteotomy. To (Fig. 7) prevent recurrence of the hallux valgus, pronation must always be managed with an arch support and a flexible shoe.

Overweight patients and patients with severe general manifestations of Down Syndrome have very severe ligamentous laxity and painful flat feet. In the absence of response to arch supports, an isolated sub-talar fusion may be performed in children under the age of 6 or the triple arthrodesis may be employed from early adolescence and throughout adult life. The isolated sub-talar (Fig. 8,9) arthrodesis may be performed according to a variety of techniques including those of Batchelor-Brown, Grice, and Dennyson-Fulford. Over the last 25 years the author has employed such surgery in only 14 out of nearly 300 patients. This suggests that careful attention to the details of conservative management, together with weight reduction and appropriate foot gear, will relieve most patients without having to resort to surgery. In the study reported to the American Academy of Orthopaedic Surgeons in 1974 and published in 1981 by Diamond, Lynne, and Sigman[6], 107 patients with 214 feet were evaluated. One hundred percent of those patients had significant pes planus. Only 10 such patients required surgical intervention for this complaint.

## KNEE

In the study cited above, 60 percent of all knees showed abnormal mobility of the patella. Twenty-two percent subluxed but did not dislocate and 5 percent were

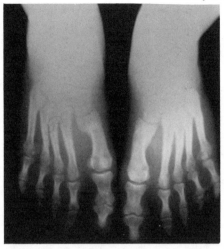

Fig. 7. Surgical correction of bunions by basal osteotomy of 1st metatarsal and capsuloplasty.

completely dislocated at initial examination. In patients without Down Syndrome severe genu valgum and lateral insertion of the infrapatellar tendon into the tibial tubercle with increased Q angle are associated with dislocation and subluxation of the patella. By contrast, in Down Syndrome neither increased Q angle nor severe genu valgum are common forerunners of dislocation. Male patients do show a higher Q angle if they go from simple patellofemoral instability to fixed

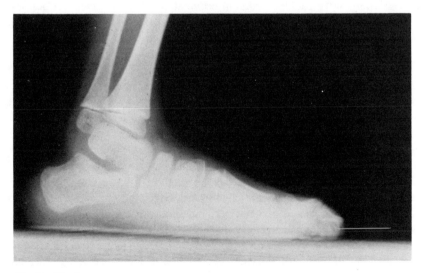

Fig. 8. Flat foot.

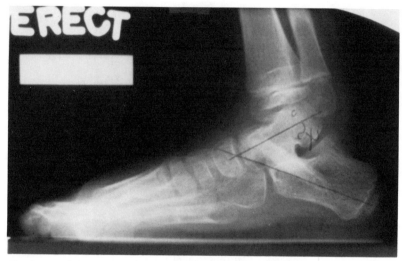

Fig. 9.   Batchelor-Brown subtalar arthrodesid for severe flat foot.

dislocation but female patients do not. This observation is best supported by the findings of Dugdale and Renshaw[7] published in 1986.

Recurrent dislocation of the patella or severe subluxation with effusion, pain, and giving way requires surgical patellofemoral reconstruction. This includes lateral retinacular release, release of the iliotibial band including the intermuscular septum, repositioning of the patella in the intercondylar grove, medial capsular reefing, advancement of the vastus medialis, and in skeletally mature patients, repositioning of the patellar tendon insertion at the tibial tubercle. If knock knee is present, the tibia valga deformity must be corrected prior to repositioning of the patella. (Fig. 10).

Patellar reconstruction should be undertaken at an early age. Long standing dislocation or chronic subluxation results in an intercondylar notch which is undeveloped and which will not receive a repositioned patella. Unless dislocation is chronic to the point of being nearly constant it is better to wait until the patella ossifies before beginning surgical management. This means the patient will be at least 6 or 7 years old and thus better able to cooperate with the prolonged rehabilitative process. Post-operatively, a cylinder cast is used for 6 weeks and bivalved. Gradual progressive motion is begun under direct and constant supervision of a physical therapist. Home therapy should not begin until knee flexion reaches 45 degrees. The patient walks in cast beginning one week post-operatively but free and unencumbered walking should not be allowed until knee flexion reaches 90 degrees and the patella tracks in the intercondylar groove over the whole range of active motion. Passive motion should never be utilized during therapy. Careful, gentle, and slowly progressive quadriceps exercises are also essential. In patients with patellofemoral instability, those athletic activities which

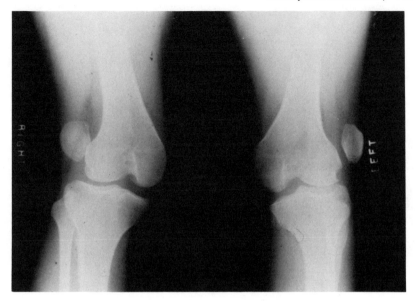

Fig. 10.    Fixed dislocation of the patellae with unstable tibio-femoral joints.

require sharp turns and cutting should be discouraged because of the torque effect. In the initial post-surgical rehabilitation period of at least 2 years, those activities should also be discouraged. On the other hand, bicycle riding, swimming, and non-torsional running can be permitted. Jumping which can produce sudden sharp flexion at the knee and kicking sports such as soccer, are not appropriate until after the first two post-operative years.

Adult patients presenting with symptomatic fixed dislocation of the patella usually have no intercondylar groove. Patellectomy is a satisfactory solution for their discomfort and disability. For those adults who have no disability, no treatment is required. Patients with unilateral dislocation of the patella and subluxation of the tibio-femoral joint, require knee fusion in order to become satisfactory and prolonged walkers. Unlike Dugdale and Renshaw,[7] we have not been successful in the use of long leg articulated orthoses in patients with grossly unstable knees. The efficiency of the orthotic is often limited by the soft and doughy consistency of the leg (Fig. 11) muscles and by the patient's acceptance of the orthosis.

On the other hand, patients with bilateral instability of the tibio-femoral joint must be managed with braces since bilateral knee fusion or fusion with an unstable opposite knee can present an insurmountable problem for an ambulatory patient. Bilateral knee fusion precludes stair climbing and unilateral fusion with an unstable knee on the opposite side places an inordinate strain on the latter joint.

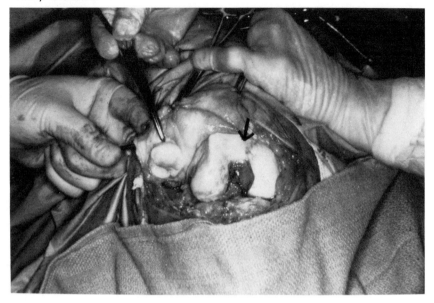

Fig. 11.   Absent intercondylar groove of femur with dislocated patella.

## HIPS

While the numbers of severely unstable hips are not great in Down Syndrome (about 10 percent), the technical challenge is worthy of the best among us. The combination of extreme laxity of joint capsules and very bizarre sitting postures produce early and late developmental subluxation or dislocation. Patients with Down Syndrome rarely have dislocated hips at birth. The hips usually remain stable but hypermobile until walking begins at 2 to 3 years of age. The pelvis has a characteristic appearance with a broad iliac crest and acetabulum which may be both well formed and deeper than that of a normal child. Because of hyperlax joints and bizarre sitting posture, the hips begin to habitually dislocate and spontaneously relocate between the ages of 2 and 10. Gradually the lateral acetabulum becomes eroded and concentric reduction becomes both less common and less possible. Occasionally an acute dislocation will occur and can easily be reduced under general anesthesia. Finally, in late adolescence, habitual dislocation and subluxation progresses to fixed dislocation. These patients are able to walk but have a pronounced limp if the lesion is unilateral and a remarkable waddle if it is bilateral. They may not have pain until mid adult life when severe degenerative arthritis occurs. (Fig. 12).

Braces and harnesses are not useful. Varus derotation osteotomy of the proximal femur correcting both coxa valga and anteversion is essential for maintenance of reduction. Snug capsuloplasties should be performed and should be protected by cast post-operatively and then by careful monitoring of the sitting position.

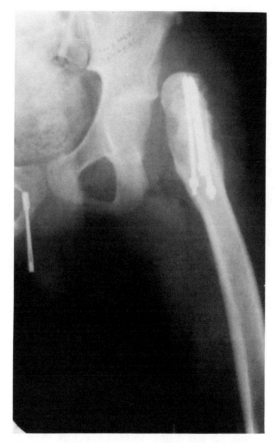

Fig. 12. Habitual dislocation of hip in patient with slipped capital epiphysis treated with multiple pins.

Where acetabular deficiency exists, Salter pelvic osteotomy is helpful to realign the socket but other types of pelvic osteotomies including the Chiari have not been effective. The largest series of patients with dislocation of the hip in Down Syndrome was presented by Aprin, Zink and Hall[1] in 1935. In their paper, they emphasized the importance of the repair and imbrication of the capsule because bony procedures alone will not retain the hip in the acetabulum in the absence of adequate soft tissue buttress. On the other hand, soft tissue procedures alone are also doomed to failure.

## SLIPPED CAPITAL FEMORAL EPIPHYSIS

A less common problem with serious sequelae is acute or chronic slipping of the proximal femoral epiphysis. (Fig. 13) Displacement of this growth center in

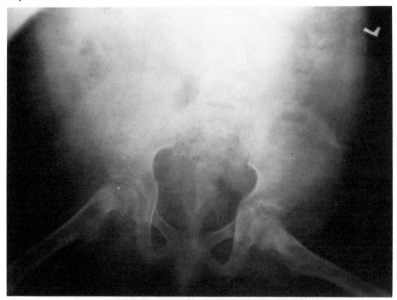

Fig. 13. Slipped capital femoral epiphysis.

adolescence results in a high incidence of both avascular necrosis of bone and cartilage necrosis. Even with careful surgical management, severe and early destruction of the hip joint is the rule. It is not yet known whether the results will be better with the new single pin surgical technique in which the surgeon assisted by a pre-operative CAT scan is able to place a single lag screw dead center across the epiphyseal line. What is known is that the prognosis for slipped capital femoral epiphysis in Down Syndrome is very much worse than the prognosis for a normal child. It is not unusual for the patient to sustain complete re-absorption of the femoral head with dislocation of the hip. In patients who are poor candidates for total joint replacement, hip fusion is the only recourse.

Because patients with slipped capital femoral epiphysis often complain of knee pain rather than hip or thigh pain, X-rays of both the hips and knees should be obtained in order to avoid missing the diagnosis. All hip X-rays should be bilateral because early unilateral changes can be very subtle and should include a frog lateral which may be diagnostic when the anteroposterior view is equivocal. A CAT scan can also be diagnostically useful, particularly the transverse sections.

## SPINE

Over the last 20 years there has been an increasing focus of physician, parental, and public attention on the disorders of the spine in Down Syndrome. Much of this interest has been generated by a desire of patients and their families to participate in the Special Olympics. Scoliosis is moderately prevalent in Down Syndrome. In the survey by Diamond, Lynne and Sigman, 52 percent of institu-

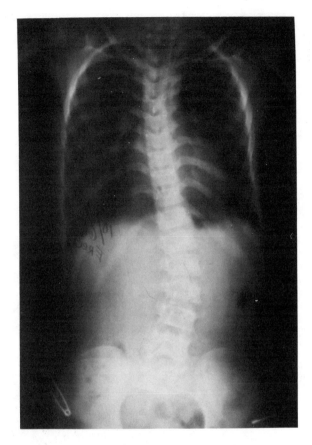

Fig. 14.   Mild scoliosis.

tionalized patients had mild curves requiring no treatment. Less that 5 percent of patients went on to develop deformities sufficient to require spinal bracing and only a rare patient required surgery. (Fig. 14).

Early degenerative changes occur in the spine in Down Syndrome. By the mid 20s most patients have significant disc space narrowing and signs of osteoarthritic changes occur throughout the spine. Most of these patients are asymptomatic or have only minimal complaints.

In the early 1970s orthopedists first recognized the unique problem of the cervical spine. Instability of the atlanto-axial joint with disconnection of the head and neck articulation and impingement on the spinal cord was observed to result in weakness, paralysis, and even death. Since 1981 in the English language, literature alone, more that 12 major papers have been published on this subject, some centering around management and others around the diagnostic question. The consensus which emerges is as follows: (Fig. 15, 16).

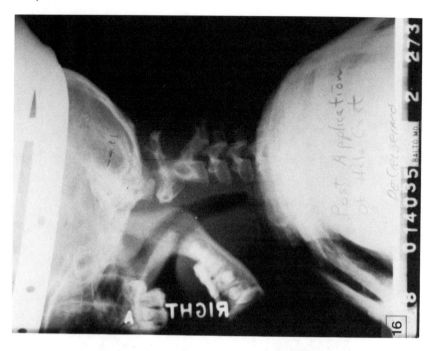

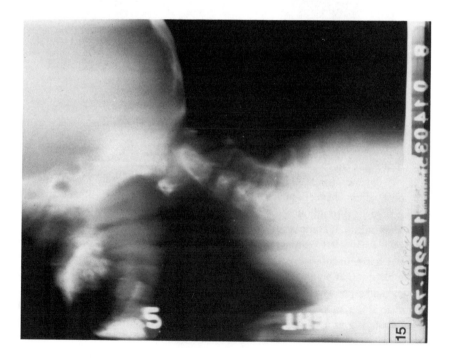

1.  The combination of ligamentous laxity, odontoid maldevelopment, and secondary odontoid damage produce varying degrees of instability at the occiput—C1-C2 articulations. Often a seemingly minor event can produce serious neurological problems at a life threatening level.

2.  It is not yet known whether cervical spine instability is static or progressive. Some authors have found slow progression into adult life and others have found the problem to be non-progressive. A rare patient may spontaneously improve. Consequently, a patient can not be evaluated once and then dismissed.

3.  In the presence of a stable spine, certain activities must still be avoided because the stability is only relative and easily disrupted. Tumbling, somersaults, trampoline, diving, and body contact sports are unsafe. Prohibited sports should include football, wrestling, high jump, karate, rugby, and sudden acceleration-deceleration activities such as roller coasters and other whip type carnival rides. Potential for gross instability under stress also makes it necessary for children and adults to travel in a high back automobile seat appropriate to body size. One 23 year old patient known to the author became quadriplegic while riding in the cargo space of a van belonging to his day care center. He was sitting on the floor with a group of other late adolescents when the van stopped suddenly. He became a permanent quadriplegic due to acute C1-2 subluxation with spinal cord impingement.

4.  Patients known to have gross instability were at special risk because of their hyperactivity or who have had neurological symptoms deserve to be treated by spinal fusion appropriate to their level and type of instability. Careful attention must be paid to the restoration of a straight and unencumbered path for the spinal cord from its exit from the skull at the occipital foramen throughout the cervical spine.

5.  Because of head position required during general anesthesia, Down Syndrome patients undergoing surgery should have a current evaluation of the cervical spine including lateral flexion and extension films, anteroposterior and odontoid views. It is important to determine whether there is an odontoid process present and whether it is in one or more pieces. One can not fully assess the status of the cervical spine under age 5 years. X-rays should be taken then at age 5 and repeated at age 10, 15, and 20 years and once again between 25

---

Fig. 15.   Unstable C1-2 articulation —tomogram. Quadriplegic.

Fig. 16.   Halo traction reduction with improvement in quadriplegic status.

and 30 years to look for possible progressive changes. The appearance of accessory ossicles at the odontoid indicate that injury has occurred. The natural history and management of cervical spine problems in Down Syndrome have been particularly well outlined in Burke[4] in 1985 and by Pueschell [14, 15] et al. in 1970 and again in 1990.

## CONCLUSIONS

The orthopedic management of Down Syndrome requires careful attention to a wide variety of skeletal manifestations. The reconstructive surgeon must be exceptionally attentive to surgical detail and rehabilitation planning and supervision. While the orthopedic problems in Down Syndrome are challenging, no patient should be refused care solely because he has Down Syndrome. Orthopedists involved with these patients should be mindful that the more ambulatory and physically functional the patient is, the more easily he can be accepted into the general community. For an institutionalized patient, ambulatory status plays an important role in intra-institutional placement and helps to determine the level of freedom and autonomy available to the individual.

## SUMMARY

### Orthopedic Disorders in Down Syndrome

When the life expectancy of patients with Down Syndrome was limited by congenital heart disease, infectious disease and tumors, very little attention was paid to the skeletal manifestations of the syndrome. With the extension in life potential, orthopedic surgeons have been compelled to explore the special rehabilitation challenges in the problems seen in the feet, knees, hips, and spine of these patients. These problems include severe flat feet, complex bunions, dislocation of the patella, dislocated hips, slipped capital femoral epiphysis, and a variety of problems in the spine. Scoliosis is not a major problem for these patients but instability of the occipito-atlanto axial articulation is a major and life-threatening problem.

Carefully chosen surgical techniques specifically designed for patients with Down Syndrome and meticulous attention to physical therapy coupled with a prolonged post-operative rehabilitation program revealed good results for most patients.

No patient should be turned away for reconstructive surgery simply because he has Down Syndrome. In addition to the surgical considerations for rehabilitation, careful preventive measures can be used to avoid some of the skeletal problems. Children with Down Syndrome should be fitted with proper shoes which will not compress the toes and thereby produce serious problems with bunions. Bizarre sitting postures should be avoided to reduce the potential for developmental dislocation of the hips and the status of the knees should be closely monitored so that recurrent dislocation of the patella can be treated before it becomes chronic and fixed.

In order to protect the very delicate cervical spine which is subject to dislocations with minimal trauma, patients with Down Syndrome should avoid somersaults, trampoline, body contact sports, diving, roller coaster and whip-like carnival rides. All patients with Down Syndrome, both children and adults, should ride in a high back automobile seat of appropriate size.

Careful attention to the skeletal problems of patients with Down Syndrome can improve their ambulatory potential with secondary improvement in lifestyle and communal interaction.

## REFERENCES

1. Asprin, H., Zink, W.P., and Hall, J.E. Management of Dislocation of the Hip in Down Syndrome. J. Ped. Orthopedics, 1985:5:428-431.
2. Bennet, G.C., Rang, M., Roye, D.P., Asprin, Dislocation of the Hip in Trisomy 21. J. Bone and Joint Surg. 1982:64-B, 289-294.
3. Blum-Hoffman, E., Rehder, H., and Langenbeck, U. Skeletal Anomalies in Trisomy 21. American Journal of Medical Genetics, 1988:29:155-160.
4. Burke, S.W., French, H.G., Roberts, J.M., Johnston, C.F., Whitecloud, T.S., and Edmunds, J.O. Chronic Altanto-Axial Instability in Down Syndrome. J. Bone and Joint Surg. 1985:67-A, 1356-1360.
5. Cope, R. and Olson, S. Abnormalities of the Cervical Spine in Down's Syndrome. Southern Medical Journal 1987:80:33-36.
6. Diamond, L.S., Lynne, D., and Sigman, B. Orthopedic Disorders in Patients with Down's Syndrome. Orthop. Clin. North America. 1981:12:57-71.
7. Dugdale, T.W. and Renshaw, T.S. Instability of the Patellofemoral Joint in Down Syndrome. J. Bone and Joint Surg. 1986:68-A:405-413.
8. Enison, G. and Oliver, M. Acetabulum Excavatum in Adult Down Syndrome. Brit. Journal of Radiology. 1981:54:340-342.
9. French, H.G., Burke, S.W., Roberts, J.M., Johnston, C.E., Whitecloud, T., and Edmunds. J.O. Upper Cervical Ossicles in Down Syndrome. J. Ped. Orthopedics. 1987:7:69-71.
10. Hreidarsson, S., Magram, G., and Singer, H. Symptomatic At lanto-Axial Dislocation in Down Syndrome. Pediatrics 1982:69:568-571.
11. Herring, J.A. and Fielding, J.W. Cervical Instablility in Down Syndrome and Juvenile Rheumatoid Arthritis. J. Ped. Orthopedics. 1982:2:205-207.
12. Livingstone, B. and Hirst, P. Orthopedic Disorders in School Children with Down's Syndrome with Special Reference to the Incidence of Joint Laxity. Clinical Orthopedics and Related Research. 1986:207:74-76.
13. Skoff, A.D. and Keggi, K. Total Hip Replacement in Down's Syndrome. Orthopedics. 1987:10:485-489.
14. Pueschel, S.M., Findley, T.W., Furia, J., Gallagher, P.L., Scola, F.H., and Pezzullo, J.C. Journal of Pediatrics. 1987:110:515-521.
15. Pueschel, S.M. and Scola, F.H. Atlanto-Axial Instability in Individuals with Down Syndrome. Epidemiologic, Radiographic, and Clinical Studies. Pediatrics. 1987:80:555-560.
16. Pueschel, S.M., Scola, F.H., Tupper, T.B., and Pezzullo, J.C. J. Ped. Orthopedics. 1990:10:607-611.
17. Sherk, H.H., Pasquariello, P.S., and Walters, W.C. Clinical Orthopaedics and Related Research. 1982:162:37-40.

18. Shikata, J., Yamamuro, T., Mikawa, Y., Hirokazu, I., and Kobori, M. Surgical Treatment of Symptomatic Atlanto-Axial Subluxation in Down's Syndrome. Clinical Orthopaedics and Related Research. 1987:220:111-118.
19. Tredwell, S.J., Newman, D.E., and Lockitch, G. Instability of the Upper Cervical Spine in Down Syndrome. J. Ped. Orthopedics. 1990:10:602-606.

# Recurrent Otitis and Sleep Obstruction in Down Syndrome

Marshall Strome, MD, and Scott Strome, MD

The otolaryngologic manifestations of Down Syndrome are not infrequently undetected because of masking anatomic and physiological parameters. Yet this spectrum of diseases can detract from an inherently diminished level of function and potentially jeopardize survival. This paper examines the pertinent associations between Down Syndrome and two pathologic entities related to the head neck: otitis and obstructive sleep apnea. Evaluation of current therapeutic modalities reinforces the need for early diagnosis and treatment so that the potential of this patient group can be fulfilled.

While our total understanding of the pathophysiology of Down Syndrome in the head and neck is not refined, specific elements of some entities are now better understood. Often presenting with common conditions related to the ear and respiratory tract, these children are predisposed to more diverse and serious sequelae than the population at large.

## EAR PATHOLOGY

The occurrence of otitis media, middle ear effusion, and hearing loss are clearly increased in the Down Syndrome (DS) population. Pathophysiology of the aforementioned has both anatomic and functional considerations, yet these diseases are often interrelated and synergistic in their potential morbidity. Early diagnosis and modern therapeutic regimens can effectively curtail disease progression, leading to improved hearing, language development and social interaction.

Variations in the size and shape of the external canal and eustachian tube afford a structural theory for the ear pathology observed in this patient group. Schwartz has documented that the pinna in newborns with DS is two standard deviations below the norm and the diameter of the external auditory canal is decreased (Schwartz, DM,Schwartz, RH, 1978). Both Schwartz and Strome have described a positive correlation between external canal stenosis and middle ear effusion, with values of 94% and 80% respectively (Strome, 1981; Schwartz, DM,

127

Schwartz, RH, 1978). With the diagnosis of effusion made more difficult by stenotic canals, the potential for an undetected conductive hearing loss is increased.

The middle ear effusion so frequent in DS may also have an anatomic derivation. Otitis media in part relates to congenital anomalies of the eustachian tube cartilage. Two of three DS patients had cartilaginous defects in their eustachian tubes in a study of temporal bone histopathology. Using pathologic criteria, the otitis media observed in this subset was more severe (Sando, Haruo, 1990). Certainly, a diminished eustachian tubal size in conjunction with other structural anomalies can predispose to stasis, ascending infection, and ultimately, recurrent otitis media.

Another study of temporal bone pathology, inferred that the presbycusis may provide a model for decreased hearing in DS (Krmpotic-Nemanic, 1970). Analysis of 2600 temporal bone sections demonstrated progressive osteogenesis along the outflow pathway of the basal spiral tract. Similar pathologic changes appear to occur in DS patients at an earlier age and with accelerated evolution, the implication being that this represents a potential pathway for sensorineural hearing deficits in this group (Krmpotic-Nemanic, 1970). Parenthetically, it is our opinion that controlling infection early decreases the onset and progression of sensorineural hearing loss in DS.

While anatomic abnormalities are integral to the pathophysiology of middle ear disease, functional considerations are also important. The hypotonicity characteristic of DS may affect tesor veli palatini muscle function, resulting in poor middle ear aeration and subsequent effusion/infection (Schwartz, DM, Schwartz, RH, 1978). Altered T-cell function may further the potential for otitis media (Scoggin, Patterson, 1982). Ossicular fixation, probably increased in DS, can augment any associated conductive hearing loss (Strome, 1981; Schwartz, DM, Schwartz, RH, 1978). Basic comprehension of anatomic and functional components of middle ear disease in DS leads to an awareness of associated hearing problems and provides insight into treatment strategies.

The reported incidence of hearing loss in the DS population varies widely, reflecting differences in the definition of normal hearing parameters, the ages of the study groups, institutionalization status, and the types of tests administered. Unfortunately, variations between reported studies diminishes the statistical significance of many potential epidemiologic trends. Currently, it is only possible to formulate general hypotheses which will require additional data for verification.

In the studies reviewed, the incidence of hearing loss ranged from 42% to 78%. (Strome, 1981; Schwartz, DM, Schwartz, RH, 1978; Keiser et al., 1981; Brooks et al., 1972; Davies, 1988). Fulton and Lloyd evaluated 79 institutionalized patients with a mean age of fifteen, and reported a hearing loss of greater than 25 dB in 42%. They concluded that the incidence of hearing loss in patients with DS was equivalent to that observed in the entire population with mental retardation (Schwartz, DM, Schwartz, RH, 1978; Brooks et al., 1972).

M Strome observed a 50% incidence of decreased hearing in noninstitutional-ized patients, ranging in age from 1.5 to 3, using a 15 dB parameter to define hearing loss. This data was consistent with Fulton and Lloyd's earlier report, and lent credence to the idea that the incidence of hearing loss was lower than suspected in younger DS patients living at home, with adequate infection control (Strome, 1981).

Several other studies, notably Brooks (1972), Krajicek (1977), and Davies (1988), have described higher incidences of hearing loss. Some of this data undoubtedly reflects an older institutionalized patient group. M Strome has reported that the incidence of cholesteatoma and otitis media are increased in this institutionalized population, producing falsely elevated levels of conductive and sensorineural loss (Strome, 1981).

Although the overall incidence of hearing loss in DS is not satisfactorily documented, it is apparent that conductive disease is the primary etiology, with an increased prevalence of sensorineural loss in older patient populations. Having previously discussed the pathophysiology of these disorders, several points re-main worthy of further discussion.

Middle ear effusion with associated otitis media is the primary cause of conductive deficits in the DS population (Strome, 1981; Schwartz, DM, Schwartz, RH, 1978; Keiser et al., 1981; Brooks et al., 1972). Although there are several potential etiologies for this type of effusion, its documented association with external canal stenosis is valuable in both diagnosis and treatment (Strome, 1981; Schwartz, DM, Schwartz, RH, 1978). Pneumo-otoscopic examination of the middle ear, the benchmark for detecting middle ear effusion, is difficult, if not impossible, in the patient with stenotic external ear canals. Knowledge of such an association with effusion mandates persistence. We recommend otomicroscopy for any Down Syndrome child whose tympanic membranes cannot be adequately visualized with routine otoscopy.

If effusion is present, M Strome has reported that placement of ventilation tubes (VT) prior to age three, complemented by antibiotic therapy relieved fluid in 50% of cases with complete resolution of conductive losses. Of the remaining group, Strome estimated that most were amenable to successful treatment, repeat tubes, tonsillectomy, and adenoidectomy, leaving only a 10% incidence of con-ductive pathology (Strome, 1981). In an older population of 32 patients who underwent VT placement for effusion, Davies described only two cases of im-proved hearing, after two years (Davies, 1988). While intergroup variations prevent direct comparison, surgical therapy appears more efficacious at an earlier age. The message seems clear. The earlier the intervention, the more likely the process is likely to resolve without sequelae.

In Brooks' study of 100 DS patients, the incidence of sensorineural loss in patients less than twenty years old was 21%, while it increased to 55% in the older patient group (Brooks et al., 1972). Similarly, in M Strome's study of DS patients under the age of three, only three cases of sensorineural loss were described

(Strome, 1981). While some of these changes may be attributable to osteoid deposition along the basal spiral tract, age and infection are important etiologic factors (Strome, 1981; Krmpotic-Nemanic, 1970). Theoretically, early infection control will partially alleviate the sensorineural deficits in the older subset of this patient population.

Knowledge of the nature and evolution of middle ear disease in the DS population allows definition of efficacious treatment strategies. It is our conviction that every DS patient should have a complete audiologic assessment, including both air and bone tympanometry and pneumo-otoscopy, prior to age 1.5. If the middle ear cannot be visualized adequately, these patients require examination under anesthesia.

The presence of middle ear effusion mandates VT placement due to the increased incidence of secondary infection and hearing loss among this population. Aggressive medical management complements this surgical therapy, and all DS patients should undergo annual audiologic examination. In the subgroup of patients who fail to respond to surgical and/or medical therapy, manifesting a hearing loss of 15 dB or greater, amplification is suggested.

Failure to recognize and correct middle ear pathology and secondary hearing loss further impairs social function (Strome, 1981; Whiteman et al., 1986). In one parental questionnaire evaluating 30 noninstitutionalized DS patients, 14 were reported to have no otitis from birth to one year of age, 5 had greater than three episodes and PE tubes placed prior to age six, and 14 had greater than three infections without PE tubes. When language ability was tested using the Basic Skills Screening Test, 100% of the tympanotomy group and 65% of the group without infection scored above the sample median, while only 18% of the group with untreated otitis had acquired this level of function (Whiteman et al., 1986). Early diagnosis and an aggressive combination of surgical and medical therapy are the cornerstones of management for middle ear pathology in this patient group, maximizing the potential for language development.

## SLEEP APNEA

Sleep apnea has an incidence of up to 50% in the DS population (Silverman, 1988). The pathophysiology has both central nervous system and obstructive components paralleling that observed in controls. Distorted symptomatology in DS, in conjunction with the insidious nature of the disease process, can obscure the diagnosis, increasing morbidity and the potential for mortality. A knowledge of the frequency and import of this entity in DS allows early intervention, thereby decreasing the incidence of complications.

Two types of obstruction, anatomic and functional, are germane to a discussion of the pathophysiology of sleep apnea in DS. Decreased palatal width, micrognathia and midfacial hypoplasia structurally impinge upon the upper airway (Strome, 1981; Loughlin et al., 1981). Macroglossia does not seem to contribute substantially to this problem, as one study found tongue size to be normal in all patients evaluated. Apparent tongue prominence was secondary to a decreased pharyngeal dimension (Ardran et al., 1972). Similarly, although adenoid and tonsilar hypertrophy can independently produce obstruction in selected instances, medialization of the lateral pharyngeal wall is more constant. Therefore, removal of only the lymphoid tissue of the oronasopharynx often failed to produce symptom resolution (Strome, 1986; Philips, Rogers, 1988).

Functional considerations are also relevant to obstruction. In a cineflourographic study of five DS children, aged 16 months to ten years, all evidenced hypopharyngeal collapse during inspiration (Loughlin et al., 1981). Levine suggests that intermittent apposition of the tongue base to the soft palate may be secondary to hypotonicity of the genioglossus muscle (Levine, Simpser, 1982).

Sleep apnea in DS can of central origin. Clark has detailed three cases of sleep apnea in DS with the pathogenesis involving both mechanical and CNS factors (Clark et al., 1980). Although several mechanisms have been implicated, an abnormal hypoxic stimulus may serve as a pathway for centrally mediated disease (Levine, Simpser, 1982). The parental history, physical examination, and a sleep study should differentiate between obstructive and central components of sleep apnea, affording appropriate intervention.

Strome notes that snoring in all positions is the "sine quo non" for obstructive sleep apnea. The parental history is the most reliable diagnostic tool as polysomnograms in DS are frequently too abbreviated to adequately assess symptom magnitude. A thorough physical examination is essential, as subtle and important structural defects can be identified, yet the presence of an anatomic abnormality obviously does not preclude co-existent functional or central pathology.

A central nervous system etiology for apnea is characterized by a lack of respiratory muscle function, whereas an exaggerated respiratory muscular effort is evidenced with obstructive disease (Levine, Simpser, 1982). Clearly, surgery has its greatest impact in patients with obstructive disease.

The senior author feels that obstructive sleep apnea in DS children is best treated by a surgical procedure designed to increase pharyngeal dimension. The latter includes removal of the tonsils and uvula, augmented by submucosal resection of muscle wedges from the lateral palate and posterior pillar with subsequent primary rotation of the residual posterior pillar and then resurfacing the entire area. In five DS patients with both upper and lower airway malformations and sleep apnea, the operation as described produced uniform resolution of symptoms (Strome, 1986). A later study by Donaldson, using a similar procedure, led to improvement in five of six patients (Donaldson, Redmond, 1988). Now, our as yet unpublished data on 14 DS patients corroborates these earlier impressions. Thirteen of the 14 patients evidenced substantive improvement in their sleep obstruction.

Although several authors have suggested that combined tonsillectomy and adenoidectomy may be curative, they are only indicated as isolated entities when marked visible hypertrophy is present, and in general should be performed in conjunction with a uvulopharyngopalatoplasty (UPP) (Silverman, 1988; Strome, 1986; Philips, Rogers, 1988; Donaldson, Redmond, 1988; Kavanagh et al., 1986). Medical control of upper respiratory infections is an essential part of the therapeutic regimen, as infection may jeopardize an already compromised airway with potentially fatal consequences (Clark, 1980).

Failure to adequately treat sleep apnea in DS produces a broad spectrum of sequelae, ranging from daytime somnolence to pulmonary hypertension. Pulmonary hypertension with cor pulmonale is the most serious long term consequence of untreated sleep apnea. Prolonged episodes of hypoxemia, hypercarbia, and acidosis result in constriction of the pulmonary vasculature, with elevated pulmonary arterial pressure. The increased demand on the right ventricle produces sequential hypertrophy, dilation, and ultimately failure (Levine, Simpser, 1982).

## SUMMARY

This paper has highlighted specific aspects of otolaryngologic disease as present in DS children. Altered anatomy, diminished intellect, and an increased propensity for infection tend to delay diagnostic and therapeutic considerations. The progression of seemingly benign disease can adversely affect social skills and occasionally can lead to major complications. Although new therapeutic strategies can maximize functional potential, effective implementation depends on early diagnosis and a knowledge of the pathologic processes.

## REFERENCES

Ardran GM, Harker P, Kemp FH (1972). Tongue size in Down's Syndrome. J. Ment. Defic. Res., 16: 160-6.

Brooks DN, Wooley H, Kanjilal GC (1972). Hearing loss and middle ear disorders in patients with Down's Syndrome (mongolism). J. Ment. Def. Res., 16: 21-9.

Clark RW, Schmidt HS, Schuller DE (1980). Sleep induced ventilatory dysfunction in Down's Syndrome. Arch. Int. Med., 140: 45-50.

Davies B. (1988). Auditory disorders in Down's Syndrome. Scand. Audiol. Suppl. 30: 65-8.

Donaldson JD, Redmond WM. (1988). Surgical managment of obstructive sleep apnea in children with Down Syndrome. J. Otolaryngol., 17: 398-403.

Kavanagh KT, Kahane JC, Kordand B (1986). Risks and benefits of adenotonsillectomy for children with Down Syndrome. Am. J. Ment. Defic., 91: 22-9.

Keiser H, Montague J, Wold D, Maune S, Pattison D (1981). Hearing loss of Down Syndrome adults. Am. J. Ment. Def., 85 (5): 467-72.

Krmpotic-Nemanic J.(1970). Down's Syndrome and presbyacusis. Lancet, 2: 670-1.

Levine OR, Simpser M (1982). Alveolar hypoventiliation and cor pulmonale associated with chronic airway obstruction in infants with Down Syndrome. Clin. Pediatr. (Phila), 21: 25-9.

Loughlin GM, Wynne JW, Victorica BE (1981). Sleep apnea as a possible cause of pulmonary hypertension in Down Syndrome. J. Pediatr., 98: 435-7.

Philips DE, Rogers JH. (1988). Down's Syndrome with lingual tonsil hypertrophy producing sleep apnea. J. Laryngol. Otol., 102: 1054-5.

Sando I, Haruo T. (1990). Otitis media in association with various congenital diseases. Ann. Otol. Rhinol. Lanryngol. Suppl. 148: 13-6.

Scoggin CH, Patterson D (1982). Down's Syndrome as a model disease. Arch. Intern. Med. 142: 462-4.

Schwartz DM, Schwartz RH. (1978). Acoustic impedence and otoscopic findings in young children with Down's Syndrome. Arch. Otolaryngol. 104: 652-6.

Silverman M (1988). Airway obstruction and sleep disruption in Down's Syndrome. Br. Med. J. [Clin. Res.], 296: 1618-9.

Strome M (1981). Down's Syndrome: a modern otorhinolaryngologic perspective. Laryngoscope, 91: 1581-94.

Strome, M. (1986). Obstructive sleep apnea in Down syndrome children: a surgical approach. Laryngoscope 96: 1340-2.

Whiteman BC, Simpson GB, Compton WC (1986). Relationship of otitis media and language impairment in adolescents with Down Syndrome. Ment. Retard. 24: 353-6.

# Oral and Dental Considerations in Down Syndrome

Edward S. Sterling, DDS

People with Down Syndrome have been the subject of more investigation and interest throughout history than any other single group in the field of developmental disabilities. Dentistry has shared this fascination with Down Syndrome, as well. As a result, there is much that has been described regarding the dental, oral and perioral findings; these include the following (in no particular order):

- significantly higher incidence of congenitally missing primary and permanent teeth
- significantly delayed eruption of primary and permanent teeth
- reduced salivary flow
- underdevelopment of the maxilla and mid-face region
- flat or prognathic facial profile
- relative tongue enlargement
- small nose and low nasal bridge
- shortened and narrowed palate with thickened lateral processes
- prominent anterior rugae
- higher incidence of over-retained primary teeth
- orthodontic problems involving anterior and/or posterior teeth
- a typical patterns of eruption, especially of the primary teeth
- unusual shape and form of both the primary and permanent teeth
- enamel defects
- higher incidence of supernumerary teeth
- lower incidence of decay (but probably only slightly)
- significantly higher and earlier incidence of destructive periodontal disease with early tooth loss
- higher incidence of bruxism (night grinding)
- thick furrowed tongue
- hypotonicity, hyperflexibility and ligament laxity
- imprecise and slowed volitional tongue movement

These findings have implications for dental treatment throughout the life span of the person with Down Syndrome. This chapter will present what is currently known and unknown in the field of clinical dentistry as it relates to people with Down Syndrome.

Since John Langdon Down first described comprehensively the characteristics seen in what is referred to as Down Syndrome, there has been a great deal of interest to study and describe the pattern of characteristics. To many people the mental image of mental retardation is Down Syndrome. In regard to the range of characteristics recognized in individuals with Down Syndrome, one could easily fill multiple pages of text with a list (Cooley, Graham, 1991).

Children with Down Syndrome are in need of the same kinds of services and health care as any child. Dental health is an integral part of a person's general health. For persons to function optimally they must not only be free of pain and disease, they need to be healthy.

Monitoring and measuring dental development provides unique opportunities to view general growth and development. Since primary teeth begin to develop at approximately four months in utero (Figure 1) and permanent teeth are still developing well into adolescent years (Figure 2), teeth offer the practitioner a view of what has occurred over almost the entire developmental period. At the clinical level, the teeth can provide signs *that* something happened, approximately when it happened, and the approximate duration of the event(s).

In an effort to present the oral findings and considerations for dental treatment of children and adults with Down Syndrome, a developmental approach will be taken. The material presented has been drawn from cited references, colleagues and personal experiences.

## AGE: 0-6 YEARS

Since primary teeth begin to form at approximately the fourth month of the pregnancy (see Figure 1), structural defects noted in the primary dentition are reflective of events which occurred during the prenatal period. In addition, because all the primary teeth do not begin development at the same time, the location of structural defects can aid in determining the timing of an insult (Gullikson, 1973).

Infants with Down Syndrome often present with hypoplasia of the dentition which can be either localized or generalized (Kamen, 1976). Given the range of congenital malformations described in Down Syndrome, it would not be surprising to see more generalized rather than localized defects in the tooth structure. These defects can range from intrinsic discolorations which are smooth to the touch of instruments to overt defects which can be easily detected with dental instruments.

Especially the defects which occur as roughened irregularities lend themselves to an easier onset of decay. The irregularities act as natural harbors for food material to collect and dental decay to begin.

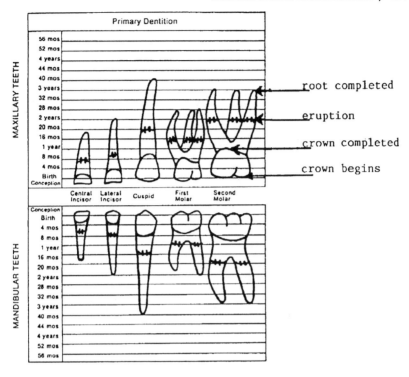

Fig. 1.   Normal development and eruption of the primary dentition.

Delayed eruption of the primary teeth is very common. In the general population, eruption of the primary teeth begins at about 6 months of age. In children
with Down Syndrome, it is not uncommon for eruption to begin at 12-14 months
of age. Clinically, the latest eruption time for a first tooth that this author has
experienced was 24 months of age.

The significance of the delayed eruption has not been determined, but it can
be a source of concern to the family. Frequently, there are questions as to
whether the child has *any* teeth. There have been no reported cases of complete
anodontia in people with Down Syndrome. Usually parental concerns can be
dealt with by simply palpating the tissues of the alveolar ridge which produces a
blanching of the gingival tissue in the shape of the underlying tooth or teeth.
Other times it is helpful to demonstrate as described and let the parent feel the
tooth through the tissue. X-rays are seldom, if ever, warranted to respond to this
request.

The prevalence of congenitally missing teeth is much higher among individuals with Down Syndrome than in the general population (Troutman, 1982).
Missing teeth occur in about 50% of individuals with Down Syndrome compared

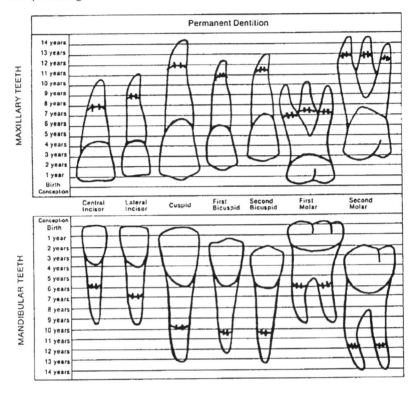

Fig. 2. Normal development and eruption of the permanent dentition (exclusive of third molars).

to approximately 2% in the general population. If the primary tooth is absent, there will be no succedaneous tooth to replace it. However, in many instances, resorption of the primary tooth will still occur and the tooth will be shed albeit somewhat later. Sometimes, the primary tooth will not be resorbed or will be resorbed so slowly that it can be retained well into adulthood. This is monitored easily through periodic X-ray.

The primary teeth of children with Down Syndrome are usually more conical in shape and the clinical crowns are frequently shorter and smaller than the general population. Clinical crowding of the teeth is not uncommon and usually effects the maxillary arch more than the mandibular arch (Gullikson, 1973; Troutman, 1982).

In children with Down Syndrome, not only is the eruption delayed, the sequence can be quite irregular. The central incisors still erupt first and the second molars usually erupt lastly; but between the beginning and the end, there is a great deal of variation. It is not uncommon for the first primary molar to follow the central incisor. Then, if it is present, the lateral incisor may erupt. When eruption

Table 1. The General Sequence of Eruption of the Primary Teeth

| Mandibular | Age |
| --- | --- |
| central incisor | 6 months |
| lateral incisor | 7 months |
| first molar | 14 months |
| canine | 16 months |
| second molar | 22 months |
| | |
| Maxillary | |
| central incisor | 7 months |
| lateral incisor | 9 months |
| first molar | 15 months |
| canine | 18 months |
| second molar | 24 months |

of the primary dentition has been completed finally, the child with Down Syndrome may be between 4 and 5 years old. The most frequently missing teeth are the lateral incisors.

As stated earlier, there is a significantly higher incidence of congenitally missing teeth. At the same time, there is a reported increased incidence of supernumerary teeth. This does not appear with nearly the same frequency as congenital absence and this author has not seen it occur in the permanent dentition. In the general population, the incidence of supernumerary primary teeth is approximately .3%.

Anterior and/or posterior crossbites may be present in the primary dentition and the classic pseudoprognathism with a Dick Tracy or Jay Leno type of profile may be evident already even in young children. Often, however, it is not so pronounced in the younger child but becomes more evident with age (Cohen, 1979; Jense, 1973). Although often described in the literature as having overly developed mandibles, the child with Down Syndrome actually has an underdeveloped maxilla and mid face. The hypoplasia of the mid face leads to the small nose and low nasal bridge associated with Down Syndrome (Librizzi, 1985).

Similarly, although frequently described as having macroglossia, the child with Down Syndrome has an average sized tongue. However, due to the hypoplastic maxilla, low tongue posture and general hypotonia exhibited, the tongue often protrudes from the mouth. This leads to an open bite in which the anterior teeth do not meet. In extreme situations, this can alter the shape of the mandible so that the problem is skeletal as well as dental (Librizzi, 1985).

The tongue is usually normal in appearance early in life, but as children with Down Syndrome grow older, hypertrophy of the vellate papillae leads to a deeply fissured or furrowed appearance of the tongue. This has often been described as a "scrotal" tongue (Troutman, 1982).

The palate is usually of normal height but frequently the palate is short and narrow with prominent anterior rugae and thickened lateral processes. This would be consistent with the hypoplasia of the maxilla (Kamen, 1976).

Bruxism is common in this population. It manifests early in life and often persists throughout the person's life. Initially, bruxism eliminates some of the secondary and tertiary grooves and fissures found in newly erupted primary teeth. By its nature then, bruxism tends to smooth the irregular chewing surfaces of the molars so they are more resistant to decay. Over time, however, bruxism can lead to overload of the supporting tissues and subsequent breakdown. In young children, however, a "transitory" bruxism is not uncommon.

There is much disagreement regarding the cause of bruxism. Many believe it is related to stress. However, in discussing a child's bruxism pattern with parents, it is usually discovered to be occurring when the child is doing nothing or engrossed in an activity. It seldom is found to be occurring at night during sleep.

In the preschool age child, bruxism rarely has required any active treatment. If active treatment were instituted, it would most likely consist of a "mouth guard" type of appliance. This does not eliminate the habit but, rather, protects the teeth. Additionally, the treatment includes redirecting the child and thereby disrupting this self stimulation activity.

The overall greatest threat to the dental health of a child with Down Syndrome is periodontal disease. Generally, periodontal disease is viewed as an adult phenomenon; however, children with Down Syndrome seem to be particularly prone to this disease (Reuland-Bosma, 1986). Some studies have reported gingival pocket formation in more than 30% of children with Down Syndrome under age 6 years (Brown, 1986).

Based on this information then, dental care should begin early for the child with Down Syndrome, usually at about the time the first teeth erupt. The aim of dental intervention at this point is primary prevention. By working with the parents or primary care givers, the dental health professional is establishing sound preventive practices in the home. These include:

- tooth brushing with or without toothpaste with a brush of proper size and type
- determination of need for fluoride supplementation
- dietary habits and recommendations

It is very common and even recommended by some dental health professionals for parents to use a washcloth to clean the teeth of young children. This author does not recommend this practice because parents, especially of children with handicapping conditions, often do not progress beyond this stage of practice. Therefore, it is recommended that a small, pediatric sized toothbrush be used. In doing so, both the child and the parents are employing an appropriate armamentarium from the outset.

Besides working with the parents or primary care givers, the pediatric dentist can provide adjunctive periodontal therapy, if needed. In selected instances, the pediatric dentist may also recommend sealants for the chewing surfaces of the primary molars.

One problem which is not unique to children with Down Syndrome but merits mention is "baby bottle syndrome" or "nursing bottle caries." The problem occurs when infants and toddlers are allowed to go to bed or nap with a bottle. Frequently, the baby bottle contains milk or apple juice. This practice can lead to severe, rampant decay even in young children before the age of two years. If parents or primary care givers continue to insist on the use of a bedtime bottle, the contents should be limited to only water. Related to this phenomenon is breast feeding on demand after the teeth erupt. Mothers who choose to breast feed on demand after an infant's teeth erupt need to provide good oral hygiene practices for the infant (Rugg-Gunn, 1985). If this does not occur, patterns of decay similar to what is seen in "nursing bottle caries" can occur.

Flossing, if the spaces between the teeth are closed, is recommended. However, if parents or care givers are experiencing difficulty with tooth brushing, adding flossing to the list of jobs to do will merely increase the sense of frustration and failure. Flossing can be approached once tooth brushing practices have been established by the adults and accepted (or at least tolerated) by the child.

## AGE: 6-15 YEARS

Above or below each primary tooth there is most often a permanent tooth developing. This development process can actually begin before birth in the general population. The mandibular first permanent molar and mandibular central incisor may begin developing before birth. Since dental eruption is so delayed for children with Down Syndrome, it is reasonable to conclude that virtually all development of the permanent dentition occurs postnatally in children with Down Syndrome. Therefore, hypoplastic defects noted involving the permanent dentition are of postnatal origin. Isolated, localized hypoplastic defects are frequently the result of significant illnesses or prolonged fevers (see Figure 2).

Similar to the eruption of the primary teeth, the eruption of the permanent teeth in children with Down Syndrome is significantly delayed, also. Whereas eruption of permanent teeth generally begins at age 6 with the emergence of the "6 year molar" or a mandibular central incisor, these same teeth may not become clinically visible until age 8 or 9. The sequence of eruption of the permanent teeth does follow a more usual pattern in children with Down Syndrome.

The permanent teeth erupt generally without any accompanying fevers or discomforts one associates with the primary dentition. In children with Down Syndrome it is not uncommon for the succedaneous tooth to erupt without the primary tooth being shed. This finding is especially true for the 6 mandibular and

6 maxillary anterior teeth—from canine to canine. In those instances, the primary teeth which are over-retained will require extraction.

The permanent teeth of children with Down Syndrome like their primary predecessors, are frequently smaller, and more conical in shape. The roots of the teeth, likewise, are shorter and thinner (Gullikson, 1973; Kamen, 1976; Librizzi, 1985).

Similar to the primary dentition, children with Down Syndrome have a much higher incidence of congenitally missing permanent teeth. The most common teeth to be missing are the maxillary lateral incisors, mandibular lateral incisors and second bicuspids. Even in those instances where the primary predecessor was present, the permanent tooth may be missing. This can be determined by x-ray and treatment decisions can then be made regarding space maintenance and/or closure.

In many articles and texts, reference is made to the decreased incidence of dental caries in persons with Down Syndrome. This is based on early research data gathered from studies conducted in institutional settings. The researchers were comparing institutionalized persons with Down Syndrome to non-institutionalized persons. In general, decay rates are lower in institutionalized populations because of the restricted diets and opportunities for snacking. With deinstitutionalization and more people with Down Syndrome never entering institutions, the finding of reduced caries rate is not nearly as significant as it was. If there is any reduction in decay rate, it is very slight (Kroll, 1970; Vigild, 1986; Ulseth, 1991).

During this stage of development, the relative prognathism, occlusal and skeletal disharmonies and deeply furrowed tongue become more evident. By this time also, bruxism which was prevalent previously will have diminished in most of the children and disappeared altogether in many. The hypotonicity, although somewhat reduced, is still evident by the usual mouth open posture and protruding tongue (Librizzi, 1985).

With the eruption of the posterior teeth, preventive measures such as sealants need to be considered on an individual basis. The role of the sealant is to fill in any irregularities in the grooves an fissures on the chewing surfaces, especially of the molars, so that decay cannot begin. Since the chewing surfaces are most prone to decay, protection against occlusal decay is a valuable service. Not all posterior teeth require sealants. The decision is based on the anatomy of the person's teeth and the judgment of the dental health professional.

Periodontal disease may be more evident in this stage of development. Compounding the situation is the occlusion; the crowding and rotation of the various teeth contribute to an already pre-existing predisposition to breakdown of supporting tissues including bone. Plaque levels do not necessarily correlate with the extent and severity of the periodontal disease. When compared to other groups with similar plaque levels, individuals with Down Syndrome developed an earlier and more extensive gingivitis (Reuland-Bosma, 1986). In addition, they exhibited more severe periodontal breakdown (Saxen, 1977).

Since delayed eruption of the teeth and occlusion play relatively minor roles in speech performance, most of the speech sound errors and irregularities noted in children with Down Syndrome are more likely due to deficits in motor performance; i.e., reflective of central rather than peripheral deficits. Related to this would be the slow and imprecise tongue movements which are frequently noted.

This is a stage in which orthodontic evaluation and treatment are likely to occur. As described earlier, anterior and/or posterior crossbites are found in almost all individuals with Down Syndrome (Kisling, 1966). The majority of posterior crossbites are bilateral and represent the largest category; anterior crossbites represent the second largest category. Orthodontic studies have demonstrated that all areas of the face and skull are deficient in persons with Down Syndrome (Shapiro, 1967; Westerman, 1975; Cohen, 1970). The relative mandibular prognathism appears to be due to a small mandible, positioned normally in relation to a highly deficient maxilla (Jensen, 1973). The person with Down Syndrome exhibits multiple growth disharmonies which is abetted by a minimal pubescent growth spurt (Librizzi, 1985). Although movement of teeth may proceed rapidly, apposition of bone appears slower thus requiring longer periods of retention or even indefinite retention to maintain the orthodontic correction.

In summary then, during this stage of development the child with Down Syndrome is entering the mixed dentition stage of dental development with a mixture of permanent and primary teeth present in the arches. The aims of dental care at this time are primary and secondary prevention. These are accomplished by:

- good home care program
- regular periodic dental check-ups at an interval determined by the needs of the child
- fluoride supplementation, if needed
- early periodontal therapy, if needed
- sealants to fill in irregular or deep grooves and fissures on posterior teeth, especially molars
- assessment of the developing occlusion to determine need and/or timing of orthodontic intervention and referral as needed

## AGE: 15 THROUGH ADULTHOOD

By this state there is little left to conjecture. The teeth, both present and absent, have been identified and decisions concerning the treatment of malocclusion have been made and steps initiated. Facial shape and appearance have been established and the child is entering adulthood.

Throughout the adult years, the single greatest problem faced by the dental health professional and the person with Down Syndrome is periodontal health/disease. It has been well established that persons with Down Syndrome have a much higher incidence of severe, destructive periodontal disease and some evidence suggest an approximate 96% incidence in persons with Down Syndrome under the age of 30. Personal experiences and observations have included people with Down Syndrome in their early and mid-thirties who undergo a rapid and generalized bone destruction with loss of tissue support and extreme tooth mobility. When this has been observed, no treatment has been successful in reversing the process. The treatment option available at that point has been extraction of all remaining teeth.

Given the rapid nature of the destruction and that the clinical findings do *not* correlate with the severity of the periodontal disease, it is evident that more than local factors are involved. When compared to other groups with similar plaque levels, individuals with Down Syndrome exhibited more extensive gingivitis (Reuland-Bosma, 1986) and more severe periodontal destruction (Saxen, 1977). Since the plaque composition is similar in the two groups, the difference may be due to an impaired host defense mechanism. Investigators have found reduced neutrophil (Izumi, 1989) and monocyte (Barkin, 1980) chemotaxis, diminished phagocytosis (Barking, 1980) and reduced bactericidal activity (Costello, 1976). Additionally, there appears to be a defect in T-cell proliferation and maturation (Wittingham, 1977; Hann, 1979). These immunologic deficiencies are currently being studied and are believed to account for the increased prevalence and severity of periodontal disease in persons with Down Syndrome.

The sequence seen often when there is breakdown of the supporting tissues is (Shaw, 1986):

- mandibular permanent incisors
- maxillary permanent incisors
- mandibular and maxillary permanent molars
- primary molars
- bicuspids
- permanent canines

Adding to the difficulty is that, due to the structural disharmonies previously described, persons with Down Syndrome are generally not good candidates for complete prostheses. As a result, the dentist is faced with the dilemma of trying to maintain and sustain a rapidly deteriorating oral environment. More frequent recall/check-up appointments, scrupulous oral hygiene and drug/chemical intervention are employed to at least slow the process.

Although the periodontal destruction is very common; it does not affect all people with Down Syndrome to the same degree. Until the mechanism of action is better understood, dental health professionals must attempt to provide as aggressive a preventive health program as possible, beginning in infancy and extending throughout the life of the person with Down Syndrome—it is indeed a life span battle.

## SUMMARY

There is much that has been written and described in reference to Downs Syndrome since John Langdon Down in 1866. Yet, there is much that is still unknown, especially in the area of clinical treatment.

To present a timetable for dental intervention, the following is offered:

### Age 0-6 Years:

- referral to pediatric dentist when first teeth erupt
- development of a preventive oral hygiene program at home including brushing with or without fluoridated toothpaste
- determination of need for fluoride supplementation
- assessment of dietary habits

### Age 6-15 Years:

- expansion of preventive oral hygiene program to incorporate flossing
- continued fluoride supplementation, if needed
- periodontal therapy, if needed
- assessment of need and placement of sealants, as determined
- evaluation of developing occlusion and orthodontic referral/treatment, as deemed appropriate

### Age 15 Through Adulthood:

- regular dental care at intervals determined by the needs of the patient
- aggressive periodontal therapy, as indicated
- prosthetic replacement of missing teeth, as deemed appropriate

This chapter has attempted to describe the oral and dental changes which occur during the life of a person with Downs Syndrome. Much has been condensed necessarily and some has been omitted unfortunately. The most glaring omission has been the community resources side of the service equation. Time and space do not permit exploration of this except to say that it is critical and that services *are* improving but still have a long way to go.

## REFERENCES

Barkin RM, Weston WL, Humbert JR, Maire F (1980). Phagocytotic function in Down's Syndrome. I. Chemotaxis. J Mental Deficiency Res 24:243-249.

Barkin RM, Weston WL, Humbert JR, Sunada K (1980). Phagocytic function in Down's Syndrome. II: Bactericidal activity and phagocytosis. J Mental Deficiency Res 24:251-256.

Reuland-Bosma W, Van Dijk JL (1986). Periodontal disease in Down's Syndrome: a review. J Clin Periodontol 13:64-73.

Reuland-Bosma W, Liem RSB, Jansen HWB, Van Dijk LJ, Der Weele L (1986). Morphological aspects of the gingiva in children with Down's Syndrome during experimental gingivitis. J Clin Periodontol 13:293-302.

Brown RH (1978). A longitudinal study of periodontal disease in Down's Syndrome. NZ Dent J 74:137-144.

Cohen NM, Arvystos MG, Baum BJ (1970). Occlusal disharmonies in trisomy-21. AJO 58:367-372.

Cooley WC, Graham JM Jr. (1991). Down's Syndrome — an update and review for the primary pediatrician. Clin Peds 30:233-253.

Costello C, Webber A (1976). White cell function in Down's Syndrome. Clin Genetics 9:603-605.

Gullikson JS (1973). Oral findings in children with Down's Syndrome. AJDC 41:293-297.

Rugg-Gunn AJ, Roberts GJ, Wright WG (1985). Effect of human milk on plaque pH in situ and enamel dissolution in vitro compared with Govine milk, lactose and sucrose. Caries Res 19:327-334.

Hann H-WL, Deacon JC, London WT (1979). Lymphocyte surface makers and serum immunoglobulin in persons with Down's Syndrome. Am J Mental Deficiency 84:245-251.

Kamen S (1976). Mental Retardation. In Nowak Aj (ed), "Dentistry for the Handicapped Patient," CV Mosby, pp 39-54.

Kisling E (1966). Cranial morphology in Down's Syndrome. Copenhagan, Munksgaard.

Kroll RG, Budnick J, Kobren A (1970). Incidence of dental caries and periodontal disease in Down's Syndrome. NY State Dent J 36:151-156.

Librizzi T, Kane JF (1985). Orthodontic considerations for the Down's Syndrome patient. In O'Donnell JP (ed), "Dental Management for Developmentally Disabled Persons in the Community," Mass Dept Public Health, pp 165-181.

Shapiro BL, Gorlin RJ, Redman RS, Bruhl HH (1967). The palate and Down's Syndrome. N Engl J Med 276:1460-1463.

Swallow JN (1964). Dental disease in children with Down's Syndrome. J Ment Def Res 8:102-118.

Troutman KC, Full CA, Bystrom EB (1982). Developmental disabilities: considerations in dental management. In Stewart RE, Barber JK, Troutman KC, Wei SHY (eds), "Pediatric Dentistry," CV Mosby pp 833-854.

Ulseth JO, Hestnes A, Slovner LJ, Storhaug K (1991). Dental caries and periodontitis in persons with Down's Syndrome. SP Care Dent 11:71-73.

Vigild M (1985). Periodontal conditions in mentally retarded children. Community Dent Oral Epidem 13:180-182.

Vigil M (1986). Dental caries experience among children with Down's Syndrome. J Ment Def Res 30:271-276.

Westerman GH, Johnson R, Cohen MM (1975). Variations of palatal dimensions in patients with Down's Syndrome. J Dent Res 54:767-771.

Wittingham S, Pitt DB, Sharma DLB, MacKay IR (1977). Stress deficiency of the T-lymphocyte system exemplified by Down's Syndrome. Lancet 1:163-166.

# Ocular Disorders in Down Syndrome

Thomas D. France, MD

Ophthalmologic manifestations in Down Syndrome have been known since the original description by Down in which he noted the prominent epicanthal folds and upward obliquity of the palpebral fissures. Since that original description there have been numerous reports on other ocular disorders, some much more important to the individuals affected. Chronic lid infections (blepharitis), lens opacities (cataracts), significant refractive errors, (myopia, astigmatism), corneal changes (keratoconus, acute corneal hydrops), ocular motor muscle imbalance (strabismus, nystagmus), iris spots or speckles (Brushfield spots), retinal changes (increased retinal vasculature), and blindness have all been described in Down syndrome.

The identification, treatment and in some cases, prevention of these ocular findings are important to preserve vision and binocular function. It is the purpose of this paper to review these ocular disorders and to describe the present methods of detection, prevention and treatment.

## PATIENTS AND METHOD

The medical records of 90 patients with Down Syndrome followed at the University of Wisconsin Pediatric Eye Clinic were reviewed and are included in this study. (Table 1) In addition, a series of patients, previously reported by Shapiro, examined locally, but not followed in the Pediatric Eye Clinic are also reviewed.

**Table 1. Study Patients**

| |
|---|
| N = 90 |
| SEX = M:F = 45:45 |
| AGE Mean = 6.4 +/- 5.9 years |
| Range = 4 months to 25 years |

## RESULTS

### Palpebral Fissures

We have previously reported our findings on the shape and size of the palpebral fissures in Down Syndrome.[1] In summary, the palpebral fissures are generally shorter than normal. We found that the fissures in Down Syndrome were less than 30 mm wide in 92% of our patients while the norm was greater than 30 mm in 83% of controls. The height of the palpebral fissure averaged 11.5 mm in both groups indicating that the fissure in Down Syndrome is relatively *rounder* than the norm.

Upward obliquity of the palpebral fissures was noted by Down in his original paper as being a unique aspect of the syndrome. Obliquity of the fissure was between 0 and 5! in 100% of the controls while it averaged 10.6! in our Down Syndrome patients. In fact, 45% had fissures that had more than 10! of obliquity.

### Iris Characteristics

Iris characteristics in Down Syndrome include a predominance of blue or grey coloration (87% versus 47% of controls) and a increased prevalence of Brushfield spots/speckles, (81% vs. 13%).

### Blepharitis

Acute and chronic inflammation of the lid margins (blepharitis) is a common occurrence in Down Syndrome patients. The etiology is not always clear but may include infection with skin organisms such as *Staph epidermitis,* or an immune reaction as seen in seborrheic dermatitis. The accumulation of fatty material at the base of the lashes may be a basis for further infection by a number of organisms. A major concern in Down Syndrome is the possible relationship of eye rubbing due to the irritation of the lids and the development of keratoconus. (See Keratoconus, below.) Treatment of blepharitis requires careful scrubbing of the lid margins daily, using a cotton-tipped applicator with soap (e.g. Johnson's Baby Shampoo®) and the use of antibiotics (e.g. sulfacetamide), if an infection is present. Anti-seborrheic shampoo should be used regularly if seborrhea is present on the scalp or face.

None of these findings are likely to cause any visual problems in our patients. There are, however, a number of significant visual problems that do occur at an increased frequency in Down Syndrome that may lead to significant disability if not recognized and treated.

The ocular findings of the patients seen in the Pediatric Eye Clinic are shown in Figure 1. It is important to remember that this is a select group seen, for the most part, because of suspected ocular problems and is not a reflection of the true prevalence of ocular problems in Down Syndrome.

### Refractive Errors

Significant refractive errors thought to be capable of reducing vision were present in 49% of the patients. The type of refractive error found is shown in

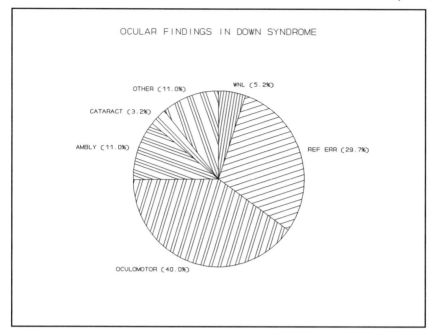

Fig. 1. Ocular disorders in patients with Down Syndrome.

Figure 2. Myopia (nearsightedness) was most common occurring in 22.6%. This is similar to that reported by Lowe (37%) in 1949 and Shapiro[1] (27% with greater than 5 diopters of myopia). Hyperopia (farsightedness), and astigmatism were present about 10% of the time for each type. The latter can significantly reduce vision for both distance and near work. Gardiner reported in 1967 that 37% of patients with Down Syndrome had astigmatism greater than two diopters and Shapiro found 25% of their patients to have astigmatism greater than three diopters. Anisometropia, a significant difference in refractive error between the two eyes, was present 7.5% of the time.

Careful refraction under adequate cycloplegia is, therefore, indicated with prescription of glasses for those cases where vision is reduced. In our experience, children of nearly any age or developmental level will happily adjust to glasses if there is a visual improvement. Those that object to their glasses need to be re-evaluated to insure that the correct refraction is in place.

### Oculomotor Imbalance

Significant oculomotor imbalance was the most common disorder seen in our clinic. (Figure 1) Figure 3 shows the breakdown of types of abnormalities seen. Esotropia (convergent strabismus) accounted for the vast majority of cases in this category. Half of these were of the accommodative type and were responsive to

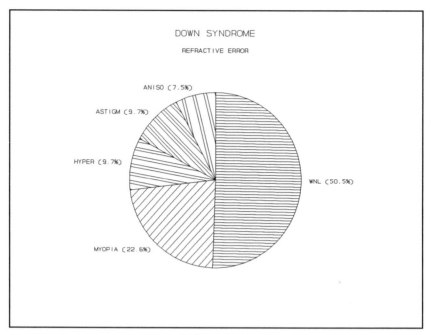

Fig. 2.  Refractive errors in patients with Down Syndrome.

correction by glasses alone. The ratio of convergent to divergent strabismus was about 10:1, or similar to that expected in an otherwise normal population of strabismic patients.

Treatment of accommodative esotropia can be accomplished by the prescription of glasses to correct significant hyperopia (farsightedness) or to reduce accommodation by the use of glasses with bifocals. Again, the children tend to accept glasses readily if binocular vision can be restored. If they are initially non-compliant, the use of a long acting cycloplegic, e.g. atropine, can be utilized to promote wearing the correction.

As in otherwise normal children, surgical correction of any residual convergent strabismus, or of divergent or vertical strabismus should be done as soon as it is well documented and vision has been equalized, (see amblyopia, below). The restoration of binocular vision at an early age is thought to allow stability of the result. Waiting until after age 5 or 6 years, may allow straightening of the eyes but binocular function may not be obtained.

Nystagmus has been found to be more prevalent in Down Syndrome than in normal controls. Sixteen per cent of our series were found to have this condition. Eissler and Longenecker, in 1962, found 15% of their patients had nystagmus and, more recently, Wagner et al. found nearly 30% of their series had nystagmus. In addition, Wagner et al. reported that 50% had vision 20/60 and 73% had esotropia.

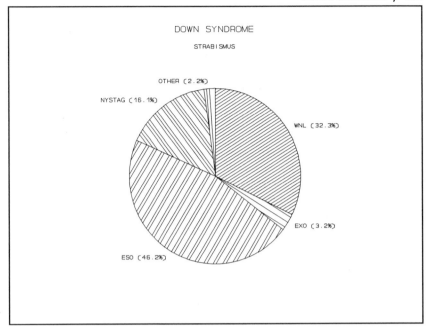

Fig. 3.   Oculomotor imbalance in patients with Down Syndrome.

The cause of the nystagmus is not known. In some cases an abnormal head position associated with the nystagmus can be improved surgically, but vision has not been improved.

### Amblyopia

Decreased vision in one or both eyes due to strabismus, anisometropia, or significant refractive errors in both eyes has been reported to be a common finding in Down Syndrome. In our series we found 11% to present with amblyopia which is more than double the expected rate. Hiles, et al. reported a series of patients with a 25% prevalence of amblyopia. In most cases these were patients with unilateral amblyopia secondary to strabismus or anisometropia.

Treatment of amblyopia includes proper correction of any significant refractive errors and occlusion of the preferred eye while monitoring the vision of both eyes. Compliance by any child to occlusion treatment is seldom good, especially when the amblyopia is severe. Encouragement of both child and parent is essential to success. Treatment is more successful and takes less time in younger children. By the age of eight or nine years, amblyopia is no longer responsive to occlusion therapy so it is important to detect its presence and initiate therapy at the youngest possible age.

## Cataracts

Cataracts have been reported to be common in Down Syndrome and are said to increase with age. Walsh has reported finding 17% and Shapiro 13% in their respective series. We found only 3% to have cataracts in our relatively young population of Down Syndrome patients and none required surgery. Others have reported the presence of lens changes in 100% of adults with Down Syndrome over the age of 35 years. Any significant lens opacity in infancy must be detected as early as possible to avoid permanent visual loss. Early surgery to remove the cataract followed by correction of the refractive error, usually with a contact lens, and then vigorous occlusion therapy to treat amblyopia are extremely important. Evaluation of the red reflex with the direct ophthalmoscope should be done during the first days/weeks of life by the primary care physician. Any significant difference between the eyes (the Brückner test), should lead to a dilated exam of the eyes and referral to an ophthalmologist if any abnormality is present. Deprivation amblyopia will develop within the first three months of life and does not respond to treatment later.

## Keratoconus

The conical disruption of the cornea (keratoconus) has been reported to be much more prevalent in Down Syndrome patients than in an otherwise normal population. Walsh found it in 8% and Shapiro in 15% of their series. Not only does this lead to decreased vision due to the effect of keratoconus on the refractive error (usually an increase in the amount of astigmatism) but it can lead to an acute breakdown in the corneal integrity resulting in acute corneal hydrops. This latter condition is characterized by pain, loss of central vision, swelling and opacification of the cornea. It is due to a break in Descemet's membrane associated with thinning of the cornea.

The etiology of keratoconus is not well established. Usually it occurs without obvious cause and has been thought to be secondary to a defect in the corneal stroma. Several authors have found a significant relationship between keratoconus and eye rubbing and atopic disease.' In addition it is known to occur more frequently in patients with Leber amaurosis when this is accompanied by the oculo-digital sign of constant eye prodding. In view of these findings, it seems wise to avoid any tendency to eye (or eyelid) rubbing. The presence of chronic blepharitis can be a definite cause of ocular irritation with subsequent rubbing, and for this reason I feel that careful attention to treatment of this seemingly mild condition should be made a high priority in children with Down Syndrome. In addition, careful attention to discourage eye rubbing at any age is equally important.

Treatment of keratoconus includes correction of the changing refractive error by glasses and/or contact lenses and ultimately by corneal transplantation. While contact lenses and corneal transplants are frequently successful in restoring vision in the normal population, such methods of treatment become much more significant if the patient is not able to fully comprehend and cooperate with the physician. The danger of traumatizing the eye following surgery is a significant concern and increases the risk of such treatment considerably. Prevention of the condition in the first place by avoiding eye rubbing is certainly best!

## Visual Function in Down Syndrome

The assessment of visual function in children with Down Syndrome is an essential part of any eye examination. Methods of visual testing in all children has changed significantly during the past few years. It is now possible to develop some understanding of the level of vision in children from a few days of life until they finally are able to perform the usual visual acuity test using the standard visual acuity (Snellen) letter chart.

These methods of visual assessment include qualitative methods such as assessment of any fixation preference between the two eyes using toys as fixation targets. More quantitative methods include matching letters, the "E" game and Allen cards, all of which allow visual assessment using Snellen-like targets. Such testing is possible in children from developmental age 2 to 3 years.

Visual assessment using preferential looking methods, as developed in the laboratory can now be used in the clinical setting using Teller Acuity Cards. This has allowed us to track the development of the visual system in infants from birth to age three years or older. The Teller Acuity test is a grating test, not a recognition test, although it is often reported in the typical Snellen equivalent of "20/20", "20/200", etc. It has been found useful in the visual assessment of handicapped children with delayed development or in those unable to communicate normally. Teller Acuity Card testing will allow us to track visual development in children with Down Syndrome as well and we should be able to show if there are significant differences in visual development in this condition as compared to the norm.

## DISCUSSION

The ocular disorders found in patients with Down Syndrome are composed of two types: those that can lead to visual loss and blindness and those that cannot. The latter may be of interest to the categorization of the syndrome but do not significantly affect the individual.

The significantly increased prevalence of refractive errors, oculomotor imbalance, especially strabismus, cataracts and keratoconus in these patients make it imperative that they be seen early by an ophthalmologist acquainted with the potential ocular findings. As with any newborn, examination by the pediatrician for anisometropia, strabismus and cataracts using the red reflex (Brückner) test should be done in the first few days of life. Careful evaluation of the refractive error and screening for strabismus should be done by an ophthalmologist by age two years so as to allow the best possible results of treatment.

Treatment of blepharitis, and continued screening for significant refractive error, cataracts, and keratoconus should be continued throughout life.

## CONCLUSIONS

Patients with Down Syndrome are at greater than normal risk for ocular disorders. Many of these disorders can lead to significant visual loss if not recognized and treated early. Conversely, if recognized and treated during childhood, such conditions as strabismus, amblyopia, and significant refractive errors can be successfully corrected. Later onset of cataracts and keratoconus should be monitored throughout life and treated as necessary.

## REFERENCES

1. Shapiro MB, France TD: Ocular manifestations of Down Syndrome, Am J Ophthalmol 1985; 99:659-663.
2. Lowe R: The eyes in mongolism. Brit J. Ophthalmol 1949; 33:131-154.
3. Gardiner PA: Visual defects in cases of Down's syndrome and in other mentally handicapped children. Brit J Ophthalmol 1967; 51:469-474.
4. Eissler R, Longenecker JP: The common eye findings in mongolism. Am J Ophthalmol 1962; 54:398-406.
5. Wagner RS, Caputo AR, Reynolds RD: Nystagmus in Down's Syndrome. Ophthalmol 1990; 97:1439-1444.
6. Hiles DA, Hoyme SH, McFarlane F: Down's syndrome and strabismus. Am Orthopt J 1974; 24:63-68.
7. Walsh SZ: Keratoconus and blindness in 469 institutionalized subjects with Down's syndrome and other causes of mental retardation. J. Ment Defic Res 1981; 25:243-251.
8. Pierse D, Eustace P: Acute keratoconus in mongols. Brit J Ophthalmol 1971; 55:50-54.
9. Ridley F: Eye-rubbing and keratoconus. Brit J Opthalmol 1961; 45:631.
10. Teller DY, McDonald MA, Preston K, Sebris SL, Dobson V: Assessment of visual acuity in infants and children: The acuity card procedure. Dev Med Child Neurol 1986; 28:779-789.

# Adults With Down Syndrome

# Sexuality and Community Living

William E. Schwab, MD

Over the past twenty years there have been dramatic improvements in medical care for children with Down Syndrome. These advances have meant that nearly every child born with Down Syndrome can now expect to live into adolescence and adulthood. Of equal significance, there has also been a dramatic evolution in the life options available to young adults with Down Syndrome during this period of time, so that individuals who in the past would be routinely confined to institutions can now expect to live and work in community-based settings. Consequently, any consideration of the health care needs of people with Down Syndrome must include a discussion of the physical changes associated with puberty, the emotional changes of adolescence, and the social changes that define contemporary community living.

## TRANSITION PLANNING

For most individuals with Down Syndrome, successful adjustment to adolescence and young adulthood requires active facilitation. This approach has come to be known as transition planning. Transition planning is an anticipatory process that should be ongoing throughout the lifespan to assure that current services are consistent with future needs. Transition planning is substantially more than just thinking about the future. Rather, it begins by collectively developing a vision of what the future might look like for an adolescent with disabilities, and then requires evaluation of the content and configuration of the supports that are being provided to assure that they are consistent with that vision. This means that educators, family members, and health care providers all need to maintain an ongoing awareness of future goals as they consider the daily decisions they make that impact on the life of a young person with Down Syndrome.

Transition planning has a number of distinct characteristics. First, transition planning is intrinsically *values-based*. It is understood that the aspirations of each individual and family are going to be determined by a combination of personal values, family values, and community values. Transition planning is also *goal-oriented*. It requires the development of a detailed plan for the future and necessitates periodic review. Finally, transition planning is *process-sensitive*. The

157

process of openly discussing values, talking about goals, sharing feelings and making changes is often just as important as writing the plan itself.

In summary, transition planning is essential for all youth with Down Syndrome. In actuality, transition planning is now something that takes place at many ages. Children who are in birth to three early intervention programs have transition plans to facilitate entry into pre-school programs. They may then have another transition plan when they move into the public school system. So transition planning is not a completely new concept, rather it is being newly applied to the period of adolescence and young adulthood.

The significance of transition planning was recognized recently in the 1990 re-authorization of Public Law 94-142, the Education of the Handicapped Act, now known as P.L. 101-476 or IDEA, the Individuals with Disabilities Education Act. In this legislation, congress defined transition planning as a "related service", comparable to physical therapy, speech therapy, and nursing services, in order to emphasize its importance. Families now have the right to expect that as part of the individualized educational plans (IEP's) for their children, they will receive assistance with long range transition planning. Health care providers should anticipate involvement in this process, especially during the child's adolescent years.

## ADOLESCENCE

Adolescence is universally recognized as a complex time of life. A number of different "tasks of adolescence" have been described. *Physical development* is the most obvious. As will be further discussed, Down Syndrome has relatively limited impact on this process. Also important is *emotional maturation*. Because individuals with Down Syndrome are likely to have led complex and atypical developmental and social lives up to the point of adolescence, the process of emotional maturation may be disrupted.

*Self-identity* is another key challenge during the adolescent years. The ability to visualize oneself as a whole person with a distinct combination of skills and strengths, as well as certain limitations, is profoundly affected by the presence of disability. This is a critical issue for youth with Down Syndrome that must be addressed in a very direct, planned, and thoughtful way. *Self-direction* is also a major task of adolescence. This relates to the ability and the right to make decisions for oneself. It means being able to make choices and to develop future plans according to personal preference. The opportunity for self-direction may be challenged by the intercurrent presence of disability as well. Other people may consistently assume decision-making roles without meaningfully involving the individual with Down Syndrome. It is essential that choices be made available in every facet of life whenever possible and, most importantly, that preferences be respected, even when there is concern about the wisdom of the decision. The right to make mistakes is a fundamental aspect of adolescence that is critical to the learning process.

Finally, adolescents begin to develop a *future perspective*. That is, they begin to mentally position themselves as adults in the world and to conceptualize what adulthood will be like for them. A component of this perspective is the realization that things that happen in the present have implications for the future. This, too, must be actively and overtly nurtured in youth with Down Syndrome.

The transition from adolescence to adulthood is typically influenced by a combination of individual characteristics, family structure, and community resources. Clearly each of these factors are affected by the presence of Down Syndrome. For example, individual characteristics—what people are good at, what they like, what skills they have—substantially influence the evolution of self-identity and self-direction. Youth with Down Syndrome require ongoing recognition and support for their abilities so that they can develop positive self-esteem and feel comfortable expressing their individuality and preferences.

Similarly, family experiences can be expected to shape and define personal expectations of adulthood. In this domain, there is unquestionably tremendous variation among families of children with Down Syndrome. Health care providers need to acknowledge the strengths that are within these families as they confront and address disability over the years, rather than viewing them as inherently dysfunctional units that are going to have tremendous problems dealing with this stage of life. Finally the community becomes a critical element in considering transition issues for youth with disabilities because the configuration of services within the community is going to very much determine what opportunities exist.

There is a concept in adolescent studies known as "launching" that refers to the process of becoming independent. Launching happens physically when a child lives away from the family home, it happens economically when a child is financially self-sufficient, and it happens in decision making when a child assumes autonomy in directing his/her life. The experience of adolescent launching can be substantially skewed by the co-presence of disability. Facilitating successful launching must be addressed in a direct way by health care providers, educators, and families within a transition planning process rather than in a haphazard manner.

The tasks of adolescence, the factors that influence transition to adulthood, and the nature of the launching process are all universal aspects of becoming an adult. As described, the presence of a disabling condition modifies this experience in both large and small ways. Often parents and health care providers struggle to figure out which of the changes they observe in young people with Down Syndrome are "typical" of adolescence and which are directly related to their disability. Parent to parent support can be of substantial assistance in sorting these questions out. The opportunity for families to link with peers who have gone through this experience is crucial to facilitating the re-definition of family roles that inevitably occurs in this period.

The concept that is perhaps most fundamental to consideration of this subject is the understanding that Down Syndrome does not mean perpetual childhood.

Young adults with Down Syndrome are not children with adult bodies, nor are they adults with children's minds. Some observers have said that it is the longing that all parents have to protect their children from the world that gets expressed in parenting children with disabilities. Most typical teenagers are able to actively separate themselves from parental control as they go through adolescence. In fact, if this fails to occur, there is felt to be reason for concern about the emotional health of the child and family. For youth with disabilities, separation is not as easy to accomplish because of the whole range of logistical, economic, physical, and psychological factors that have been described. The result can be that childhood can be inadvertently maintained. It is ultimately the responsibility of every person involved in the lives of youth with Down Syndrome to actively support and assist them in growing up and achieving progressively increasing personal independence commensurate with their individual aspirations and abilities, no matter how limited they may seem to be.

In planning for adult needs there are many aspects of life that need to be considered. A complete examination of economic, residential, vocational, educational, and recreational issues is beyond the scope of this chapter. Instead, the remainder of this discussion will focus on the issues of sexuality, mental health and medical care.

## SEXUALITY

Human sexuality encompasses an individual's self-esteem, interpersonal relationships, and social experiences relating to dating, marriage and the physical aspects of sex. Sexuality does not just refer to intercourse, rather it involves the whole range of human interactions. Many of the psychosocial factors that impact on the sexual development of youth with Down Syndrome have already been addressed. It is also necessary for health care providers to understand the physical characteristics of adolescence in Down Syndrome.

In women with Down Syndrome, the timing of puberty and menarche are generally comparable to the general population. Approximately 70% of women with Down Syndrome ovulate and are fertile at some time during their reproductive years, with more than one-third maintaining normal cycles throughout young and mid-adulthood. This means that contraception needs to be a consideration if pregnancy is not desired.

There are no forms of contraception that are contraindicated in individuals with Down Syndrome, but as with all people, contraceptive choices must be individualized to personal preference and ability. Barrier methods that require use with every sexual contact may be more difficult for people with Down Syndrome. However, many women are able to successfully use oral contraceptive agents, either independently or with supervised administration. Other less user dependent forms of birth control may also be considered including sterilization, though this may be limited by legal constraints in some states, and intrauterine devices, though these should generally be avoided in women who have not yet had children or who would have difficulty reporting early signs of pelvic infection. There is now increased interest in the use of subcutaneous progestin (Norplant), administered by implantable, slow-release hormonal sticks that are inserted under the skin in a simple minor surgical procedure. This method conveys effective, continuous contraception for up to five years with relatively few side effects, the most common being menstrual irregularity. Disease transmission through intercourse is a related area of concern, so proper condom use must also be discussed with all sexually active women and men.

There are limited data available about the offspring of women with Down Syndrome, but the studies that have been published suggest that one-third to one-half of their children will have either Down Syndrome or other developmental problems. Much better data are needed to understand whether the etiology of these developmental problems is genetic or social.

Finally, the observation is that the onset of menopause is quite variable. Again, there are no good data about this. It has been suggested that women who experience premature aging in other dimensions of their physiology will also have premature menopause. Other women with Down Syndrome can be expected to continue menstruation into their forties.

In males, puberty may be slightly delayed compared to the general population and there is some controversy as to whether full sexual maturation occurs. While one study found that penile size and testicular volume are slightly decreased, another indicated that these parameters are equivalent to the general population. It is clear that male fertility is significantly reduced. However, past literature that indicates that no man with Down Syndrome has been known to have fathered a child has been proven incorrect by a well-documented case report of paternity published in 1989. It is likely that with greater access to community living and greater opportunities for sexual activity, there will be more case reports of men with Down Syndrome fathering children.

Given the context of the reproductive physiology and the emotional development of people with Down Syndrome, a fundamental social and ethical question must be considered: Does the presence Down Syndrome mean compulsory virginity? If the answer is yes, that every person with Down Syndrome should be celibate, then the entire direction of medical care, educational activities, social programming and family interactions around sexuality must be oriented to the prevention of sexual interactions, though the presence of sexual feelings and the innate need for personal intimacy obviously cannot be eliminated. This approach, however, will substantially distort development from adolescence to adulthood.

If, on the other hand, the answer is no, that Down Syndrome does not mean compulsory virginity, that it is acceptable for people with mental retardation and other disabilities to be sexually involved, then a different set of questions arise. Under what circumstances can sexual contact take place? What knowledge level about the physical and emotional aspects of sex is required? How will education take place? Will individuals truly be able to give consent? Should marriage be an option? What about the possibility of exploitation? What active steps should be undertaken to facilitate sexual activity and opportunity for people whose lives are much more structured and monitored than most other people's lives are?

The presence of Down Syndrome impacts on these issues in a variety of ways. First, in the interpretation of community and family values, which is how awareness of sexuality begins. One of the consequences of cognitive impairment may be some difficulty understanding differences between sexuality as it is depicted in the media and sexual activity as it occurs in real life. This is certainly a difficulty many people without disabilities experience. The same may be true of understanding family values. Most families are not very overt in talking about sex. Values and information on this subject are generally discerned in very subtle ways. These subtle messages may not be accurately interpreted in the ways they are intended by people who have cognitive impairments. Sex education programs that preferably include the participation of parents with their teenage children are clearly necessary so that accurate information can be conveyed. Fortunately, a number of excellent sex education curricula that are sensitive to the needs and abilities of people with disabilities have been developed, but their availability is limited in local communities. This means that health care providers must both assist families in addressing sexual issues themselves and also serve as advocates for making educational resources more widely available.

The configuration of community activities and services is another area that dramatically affects sexual development. It is likely to be very difficult for people with Down Syndrome to initiate intimate relationships in the context of highly structured, closely supervised community programs. Often these situations lead to "inappropriate" sexual expression, because there is simply no opportunity for "appropriate" interaction. While less structured community living arrangements in which there are greater opportunities to develop close personal relationships can result in more positive sexual expression, it is also important to acknowledge that one of the outcomes in these settings might be the existence of non-nurturing relationships and unwanted pregnancy as in society in general. These possibilities need to be openly acknowledged so that there is a planned response when they occur.

Sexual vulnerability is a major concern for families. In the legal codes of most states, there exists a dynamic tension between protecting vulnerable adults on the one hand and avoiding unwarranted restriction of personal freedom on the other. In some states it is illegal to have intercourse with someone who is mentally retarded. While this type of policy is intended to prevent people who are in power relationships or supervisory roles from sexually exploiting or assaulting individuals who may not have total comprehension of the implications of the activity, it also essentially mandates that all people with cognitive impairments be non-sexual. Other states permit sexual contact as long as the individual can consent, a legal doctrine that can result in substantial differences in interpretation and application. Regardless of the prevailing legal standard, an essential component of sex education for all people is teaching protective behaviors. For people who have cognitive impairments, the focus in most curricula is on teaching students to recognize differences in the meaning of touch from one person to another based on the context of the relationship as well as on providing strategies for responding to unwanted physical contact.

The legal issues around protection of vulnerable adults versus preservation of personal choice are addressed in other aspects of the law as well. In some states it is illegal to sterilize an individual who is mentally retarded because of concerns about the ability to give consent. While it is sometimes possible to circumvent these regulations when they exist, it is imperative that health care providers be aware of their state's regulations.

## MENTAL HEALTH

Sexuality is one issue in community living that has substantial implications for health care professionals. Another that frequently arises in medical settings is assessment of behavioral change. Caregivers may be worried about the possibility of pain, seizures, or other physical etiologies for the changes they have noted. It is important to evaluate possible medical causes with an open mind. If the patient has limited verbal ability, obtaining precise historical data may be problematic, and many conditions cannot be readily diagnosed by physical examination or laboratory testing. Health care providers and caregivers need to have substantial flexibility and engage in creative problem-solving together to assure themselves that an individual is not experiencing headaches, abdominal discomfort, or other more silent conditions.

Beyond just considering possible physiological problems when behavioral change occurs, it is critical in the medical assessment to obtain detailed information about events and changes in daily activities that affect emotional health. Behavioral change may be a communication equivalent through which a person is indicating emotional distress. It would be inappropriate to pursue a broad diagnostic work-up in search of a medical problem when the history suggests new sources of life stress.

A component of the evaluation should be consideration of coping skills. Health care providers need to ask about such things as how their patients with Down Syndrome express feelings, whether they have the opportunity to gain emotional support from close friends and family, and what leisure and recreational activities they are involved in. Psychiatric diagnoses should also be considered. People with Down Syndrome can experience depression, schizophrenia, or other forms of mental illness, though it may be necessary to modify the DSM-IIIR criteria to account for variability in cognitive ability and functional status.

There are some specific steps that health care providers can take to assist in the resolution of behavioral change beyond treating possible physical causes. Supporting coping ability by recommending stress reduction activities can be an important intervention. Working with community agencies on environmental modification, which may include changes in personnel, activities, or other components of vocational, educational, recreational, and residential programming may be undertaken. This may be difficult to accomplish unless practitioners have become closely involved with the community agencies that serve their patients.

Medication should be reserved for very clearly defined indications, and it is essential that efficacy always be assessed by pre-determined criteria. A behavioral baseline should be established in advance and the outcomes that indicate successful treatment should also be defined prospectively. Referral to behavioral and mental health specialists should be considered, though practitioners need to be realistic about the fact that many mental health professionals have little experience working with people with disabilities.

## MEDICAL CARE

Community living also requires careful planning to assure that health and medical needs are met. "Health", in this context, refers to general well-being as opposed to "medical", which relates to physician care. Health concerns include such areas as nutrition, hygiene, exercise and safety. If, for example, educational programs emphasize ordering fast-food restaurant meals as an important activity of daily living, it is not surprising that when individuals with Down Syndrome have more independent choices about food selection, they may pick high calorie, high fat menus. Job and occupational safety are also important considerations since often environments in which individuals with disabilities have the opportunity to work are environments where safety is not a prime consideration and monitoring may be lax.

Access to medical care must be actively considered in the transition from adolescence to adulthood. Individuals who have been seeing Pediatricians for their care during childhood must at some point transfer to a Family Physician or Internist for primary care and must similarly do so within specialty practices unless model programs offering lifespan care for individuals with Down Syndrome exist in their area. Accomplishing this transition requires more than just making an appointment with a different doctor. It needs to be a very pro-active process that includes personal contact between physicians to talk about the particular medical issues related to Down Syndrome as well as about the individual patient and family. It may also include discussion of general issues related to caring for individuals with disabilities in their practices. If hospitalization is anticipated, pre-planning is critical as well because nursing staffs in most institutions are likely to have had limited experience with individuals with disabilities.

Adults with Down Syndrome require the same routine periodic health screening and maintenance care that is recommended for other adults of the same age, as well as special attention to conditions related to Down Syndrome like thyroid testing, vision testing, and hearing testing as discussed in other chapters. In considering health maintenance care, there may be some important ethical challenges. For example, should participation in health maintenance procedures such as pelvic examinations be compelled even if it requires physically restraining the patient? The alternative to doing so in this example would be to accept an increase in the incidence of cervical cancer in women with mental retardation who refuse gynecological care because of their limited ability to understand the reasons for the procedure.

The solution to this and similar dilemmas is awareness of practices that are flexible and creative so that health maintenance care can be provided in a sensitive manner. One such option is an approach to the pelvic examination for women who are cognitively impaired using a protocol in which the woman is in the frogleg rather than the lithotomy position, and the clinician palpates the vagina and cervix using a single digit. The Pap smear specimen is obtained by touch using the examining finger for guidance and a gentle bimanual assessment is then performed as tolerated. This procedure is satisfactory for most routine situations and is an excellent example of a strategy for providing preventive health care in accordance with universally accepted standards in a manner that is more likely to be acceptable to the patient. Similar modifications can be utilized for other aspects of medical assessment.

## CONCLUSION

With effective transition planning, many youth with Down Syndrome can develop the skills needed to achieve substantial degrees of independence as they enter adulthood. Health care providers should expect to be active participants in this process with particular emphasis on their roles in promoting emotional maturation, addressing aspects of sexuality, and assuring the availability of quality health and medical care. Caring for adolescents and young adults with Down Syndrome in the 1990s may require re-exploration of previously held assumptions about the content of community living for people with disabilities. By collaborating closely with family members, other community professionals, and most importantly, directly with their patients who have Down Syndrome, health care providers can play a critical role in supporting full inclusion for people with Down Syndrome in all aspects of community life.

## REFERENCES

Bovicelli L, Orsini LF, Rizzo N, Montacuti V, Bacchetta M (1982). Reproduction in Down Syndrome. Obstetrics & Gynecology 59(6)(supplement): 13S-17S.

Elkins TE, McNeeley SG, Rosen D, Heaton C, Sorg C, DeLancey JOL, Kope S (1988). A clinical observation of a program to accomplish pelvic exams in difficult-to-manage patients with mental retardation. Adolesc Pediatr Gynecol 1: 195-198.

Fletcher R, Menolascino F (eds.) (1989). "Mental Retardation and Mental Illness: Assessment, Treatment, and Service for the Dually Diagnosed." Lexington, MA: Lexington Books.

Heighway S, Kidd Webster S, Shaw M (eds.) (1988). "STARS: Skills Training for Assertiveness, Relationship-Building, and Sexual Awareness." Madison, WI: Waisman Center.

Patterson PM (1991). "Doubly Silenced: Sexuality, Sexual Abuse, and People with Developmental Disabilities." Madison, WI: Wisconsin Council on Developmental Disabilities.

Pueschel SM (1987). Health concerns in persons with Down Syndrome. In: Pueschel SM, Tingey C, Rynders JE, Crocker AC, Crutcher DM (eds.), "New Perspectives on Down Syndrome," Baltimore: Paul H. Brookes Publishing Co., pp. 113-134.

Pueschel SM, Orson JM, Boylan JM, Pezzullo JC (1985). Adolescent development in males with Down Syndrome. AJDC 139: 236-238.

Sheridan R, Llerena J, Matkins S, Debenham P, Cawood A, Bobrow M (1989). Fertility in a male with trisomy 21. Journal of Medical Genetics 26: 294-298.

Sobsey D, Gray S, Wells D, Pyper D, Reimer-Heck B (eds.) (1991). "Disability, Sexuality, and Abuse: An Annotated Bibliography." Baltimore: Paul H. Brooks Publishing Co.

Summers JA (ed.) (1986). "The Right to Grow Up." Baltimore: Paul H. Brookes Publishing Co.

Turnbull HR, Turnbull AP, Bronicki GJ, Summers JA, Roeder-Gordon C (eds.) (1989). "Disability and the Family: A Guide to Decisions for Adulthood." Baltimore: Paul H. Brookes Publishing Co.

# Aging and Alzheimer's Disease in People With Down Syndrome

Krystyna E. Wisniewski, A. Lewis Hill, and Henry M. Wisniewski

The clinical signs that are often associated with normal aging, including motor, sensory, skeletal, and skin changes, have been described as occurring prematurely in adults with Down syndrome (Oliver and Holland, 1986). Similarly, behavioral and neuropsychological studies have demonstrated declines in cognitive, adaptive skills (dementia), and social functioning in individuals with Down syndrome (Fenner et al., 1987; Thase et al., 1984; Wisniewski et al., 1978, 1985a,c) especially after the age of 50 years (Zigman et al., 1987). Dementing individuals with Down syndrome may demonstrate a gradual decline in initiative and creative imagination as well as a narrowing of interests and an increase in egocentricity. These may lead to personality changes, e.g., paranoid ideations, generalized anxiety, depression, or a feeling of insecurity and inadequacy (Wisniewski and Hill, 1985b). It is not unusual for some elderly individuals with Down syndrome to become irritable and stubborn and to display enhanced symptoms of unresolved intrapersonal and interpersonal problems. The age at which declines in functioning in older people with Down syndrome can be demonstrated, however, may be related to the specificity of the examination tests used. Studies that have used behavioral scales have tended to show the onset of decline in functioning to begin at a later age than those studies that have directly examined cognitive functioning (Wisniewski et al., 1985a).

Premature aging and dementia among people with Down syndrome has been recognized since a report by Fraser and Mitchell in 1876. Until recently, however, accelerated aging among this population was not considered to be an important issue affecting major program development, because typically, the majority of people with Down syndrome were not expected to live past adolescence or young adulthood (Thase, 1982, 1984).

In the last few decades, advances in medicine and changing attitudes towards the care of mentally retarded people have prolonged the life span of people with Down syndrome as well as increased their level of cognitive functioning (Wisniewski et al., 1988). The average current life expectancy is 30 years for the

80% of people with Down syndrome who do not have congenital heart defects (Baird and Sadovnick, 1987); 25% of this population is now expected to survive past the age of 50 years (Thase, 1982). Therefore, understanding the course of brain development over the entire life span of individuals with Down syndrome and describing any declines in functioning among this population that reflect a normal course of aging have now become important applied research goals.

People with Down syndrome are at particularly high risk for Alzheimer's disease, a progressive dementing disorder with characteristic clinical signs and brain pathology (Mortimer, 1983; Wisniewski et al., 1985a,c; Henderson, 1986; Mann et al., 1990). The presence of Alzheimer-type pathology in the brains of almost all people with Down syndrome 30 years of age and older has been the cornerstone for the widely held view that people with Down syndrome will develop Alzheimer's disease not only at a much younger age (20 years earlier than the general population), but also in much larger numbers (two to five times more) than people without Down syndrome (Malamud, 1972; Burger et al., 1973; Whalley, 1982; Wisniewski et al., 1978, 1985a-c). Because of this relationship between Down syndrome and Alzheimer's disease, it has been widely assumed that knowledge about almost any aspect of one of these conditions will illuminate the other (Mann et al., 1984, 1985a,b, 1987a,b; Mann et al., 1990; Wisniewski 1990).

Advances in genetics and neurobiology have increased our understanding of the mechanisms of neurological dysfunction (e.g., different degrees of mental retardation, accelerated aging, Alzheimer's disease in Down syndrome individuals (Scott et al., 1982; Coyle et al., 1986; Epstein, 1986; Wisniewski et al., 1985a,c; Wisniewski and Hill, 1985b; Wisniewski et al., 1987). It would seem that abnormal gene products of the extra chromosome 21 would confer a unique pathophysiology on Down syndrome brains. Brain development in individuals with trisomy 21 is disturbed, showing minor and major anomalies that cause the impairment of central nervous system maturation, differentiation and function that is observed at various stages of life. It is hypothesized that an abnormal gene dosage in this aneuploidy syndrome is responsible for this long lasting process (Opitz and Gilbert-Barness, 1990). The neurological and neuropathological changes occurring in individuals with Down syndrome are found during their whole life and therefore will be described separately.

## DELAYED AND ABNORMAL CLINICAL AND NEUROPATHOLOGICAL (BRAIN) DEVELOPMENT IN INDIVIDUALS WITH DOWN SYNDROME

All individuals with Down syndrome have developmental disabilities with developmental delay (e.g., speech abnormalities, different degrees of cognitive dysfunction, hypotonia, impairment of fine and gross motor coordination). It is assumed that cognitive dysfunction in Down syndrome individuals is secondary to prenatal and early postnatal brain abnormalities and environmental factors. Similarly, it is certain that the accelerated aging and Alzheimer's disease found later in life is related to the premorbid abnormalities of gene and brain structures and functions, but the critical variables are not yet known (Coyle et al., 1986; Rabe et al., 1990).

The brains of people with Down syndrome are smaller than those of non-retarded people (Crome and Stern, 1972; Wisniewski, 1990). The hippocampus appears to be reduced by as much as 30 to 50% (Sylvester, 1983). The differences in brain sizes relative to normal are smallest after birth and increase during infancy and early childhood (Sylvester, 1986), suggesting defects in the developmental program and differentiation. The shape of the Down syndrome brain is roundish, largely because of the reduced frontal pole and the superior temporal gyrus (Crome and Stern, 1972; Wisniewski, 1990). Myelination of cerebral white matter is delayed (Wisniewski et al., 1986b), and basal ganglia calcification is most often present (Wisniewski et al., 1982). The anterior commissure is considerably (50% or more) smaller than normal in many (but not all) Down syndrome brains (Sylvester, 1986).

A number of cytoarchitectonic and synaptic abnormalities are found by light and electron microscopy studies, most of which suggest arrested and/or delayed differentiation of development (Ross et al., 1984; Wisniewski et al., 1984, 1986a). The cerebral cortex of the Down syndrome brain has a reduced number of neurons, which is particularly marked in the number of granule cells of cortical layers II and IV of the cortex (Ross et al., 1984; Wisniewski et al., 1986a; Wisniewski, 1990). Precocious cessation in the development of dendritic spines in the visual cortex has also been reported (Takashima et al., 1983).

There is no consistent agreement among investigators on whether fetal Down syndrome brains are less developed than normal brains of the same gestational age. For example, one researcher found the hippocampus in a Down syndrome fetus to be not only smaller but also less mature than that of an age-matched normal fetus (Sylvester, 1983). In contrast, another investigator found, without morphometric studies, the brains of 15- to 22-week-old Down syndrome fetuses to be similarly developed as those of normal fetuses (Schmidt-Sidor et al., 1990). They suspected that delay in neuronal differentiation most likely occurs more dramatically in the last trimester, but future morphometric studies are needed to confirm the suggestion. They also think that the abnormal gene products that regulate brain differentiation in individuals with Down syndrome may already start at the time of conception.

Another consistent microscopic finding in Down syndrome is dysgenesis of dendritic spines. In addition to a reduction in the number of dendritic spines, the spines tend to be longer and to have thinner necks (Marin-Padilla, 1976; Suetsugu and Mehraein, 1980; Wisniewski et al., 1986a). This spine morphology is normal prenatally, but post-natally, they become less differentiated. The long-necked dendritic spines persist after birth, suggesting a dysgenetic process (Takashima et al., 1983; Wisniewski et al., 1984, 1986a; Scott et al., 1983), or they undergo trans-synaptic degeneration (Marin-Padilla, 1976; Suetsugu and Mehraein,

1980). Similar spine dysmorphology is seen in other categories of mental retardation (Purpura, 1974).

Also, it is known that the neuronal membranes in Down syndrome are abnormally hyperexcitable. Scott et al. (1982) found that cultured dorsal-root ganglion neurons from Down syndrome fetuses displayed several hyperexcitable membrane properties, including decreased action potentials after hyperpolarizations and decreased voltage thresholds for action potential generation. These changes can be explained by alterations in membrane ionic channels and neurotransmitter abnormalities (Coyle et al., 1986; Mann et al., 1987a,b).

## DEFINITION OF DEMENTIA

Dementia is a symptomatological diagnosis in which there is a loss of intellectual functioning severe enough to interfere with occupational or social functioning (American Psychiatric Association, 1987). It is a statement of current functioning being lower than previous functioning levels and is consistently progressive, with a slow or rapid time course lasting in trainable individuals with Down syndrome from two to 10 years (Wisniewski et al., 1985a,c). The clinical diagnosis of dementia is difficult to make; in individuals with Down syndrome, it necessitates an intensive diagnostic workup and long follow-up.

Katzman (1981) suggests that there are about 50 different causes of dementia in nonretarded adults; in our judgment, this number of causes of dementia may also apply to individuals with mental retardation. Katzman has classified these causes into primary degenerative, vascular, and secondary dementias. Some of the etiological classifications of these dementias were summarized in our previous publication (Wisniewski and Hill, 1985b).

Premature regression in memory, temporal orientation, and other capacities can be evaluated only in the group of patients with Down syndrome relatively high IQ values (Lai and Williams, 1987). In more profoundly mentally retarded patients, we more often observe a decline in everyday activities, loss of motivation, decrease in social interaction, and apathy (Dalton et al., 1977, 1984; Zigman et al., 1987). Seizures have been reported as a possible distinguishing initial feature (Lott, 1982). Other abnormalities on neurological examination may be frontal release signs and impairment of fine and gross motor coordination (Wisniewski et al. 1978, 1985a,c). Demented individuals regress from ambulatory to nonambulatory periods in later stages and may have coma virgin, extrapyramidal signs and urinary incontinence. They finally stay in the vegetative stage for some time without any communication with the surrounding world.

## ASSESSMENT OF DEMENTIA

Any aspect of clinical dementia must be assessed within the context of background functioning. The individual's history of premorbid functioning is carefully analyzed. Thus, a drop in functioning on a standardized test, such as an intelligence test, does not per se suggest dementia. One must also take into account the degree of lowered functioning as well as the areas in which the functioning level has decreased.

Decrease in functioning level of the first of these areas, memory of the recent past or short term memory, is one of the most pronounced clinical features of dementia (Lezak, 1976). Allison (1961) has categorized the main memory deficits in non-mentally retarded individuals into four types: 1) inability to remember proper names and names of objects when the object or person is not present 2) time disorientation, which may be observed in altered sequencing of remembered events 3) topographical deficits in which spatial relationships are not remembered and 4) "amnestic indifference," in which memory is not used to aid orientation or thinking.

The second area of intellectual decline among non-mentally retarded persons is the loss of the ability for abstract thinking. This decline has been reported to be a very sensitive indicator; some investigators believe it may be the earliest indicator of dementia (Bilash and Zubeck, 1960; Clark, 1960). Among mentally retarded individuals, this decline can be demonstrated only among individuals who are considered to be higher functioning (e.g., mildly retarded).

Williams (1970) reported a third area of mental inflexibility that is demonstrated by an inability to change mental sets and by difficulties adapting to new situations or solving novel problems (Goldstein and Shelly, 1975; Reitan, 1967). The fourth area is a general slowing of activity, which can be easily observed in both non-mental retardation and mental retardation populations. This behavioral slowing affects perceptual, cognitive, and psychomotor tasks (Birren, 1963; Jarvik, 1975).

## THE USE OF PSYCHOMETRICS AND MENTAL STATUS EVALUATION TO ASSESS DEMENTIA

Psychological testing for dementia of individuals with mental retardation generally begins with an IQ test. The choice of the particular instrument depends on the estimated level of cognitive functioning as well as the history of previous testing. For learning disabled and mildly mentally retarded individuals, almost all of the usual psychometric batteries can be employed to assess current functioning levels, and the major memory deficit can be easily identified. Difficulties occur when lower functioning individuals are tested. For instance, the Leiter International Performance Scale may be more appropriate for lower functioning as well as deaf or some physically impaired individuals. Although the Slosson Intelligence Test for Children and Adults and the Stanford-Binet Intelligence Scale can be used to assess very low functioning individuals, they both rely heavily on verbal skills and, therefore, may be less appropriate than other tests (such as the Cattell Infant Intelligence Scales) for individuals at the severe and profound retardation levels.

In addition to establishing and comparing IQ scores, specific measures of attention, memory, language, gross and fine motor coordination, constructional abilities, social functioning, and personality should be taken. A great number of tests of these areas are available, but many are inappropriate for individuals with MR, particularly those at the lower levels of functioning.

Evaluating such abilities as attention and concentration of individuals with dementia is particularly difficult, and it may be necessary to interview "significant others" several times to obtain the proper information. Informant scales, such as the Vineland Adaptive Behavior Scales, the Minnesota Developmental Programming System's Behavior Scales, or the American Association on Mental Deficiency's Adaptive Behavior Scales, can be particularly useful in determining social and behavioral functioning. Results of the present testing can be particularly useful if they can be compared with those of previous testings. If not, they will serve as a baseline for retesting.

Whereas standardized psychometric evaluations provide important information, they are not fully satisfactory for all mentally retarded individuals and do not measure all areas of cognitive functioning. To supplement these tests, new tasks, derived from experimental research, and a mental status evaluation are needed. Each of these tools must be adjusted to the overall functioning level of the individual.

One example of a new test for dementia that is based on experimental research and is appropriate to mentally retarded persons, is the one developed by Dalton (Dalton, Crapper, and Schlotterer, 1974; Dalton and Crapper, 1977; Dalton and Crapper McLachlan, 1984). Employing a two-choice, matching-to-sample, and delayed matching-to-sample task, using a computer to test higher functioning, institutionalized MR persons employed by sheltered workshops. Dalton and his colleagues have been able to provide early identification and follow-up of dementing Down syndrome individuals. This paradigm has been extended by the present authors to include a more complicated three-choice, delayed matching-to-sample task using a computer in order to test higher functioning, noninstitutionalized mentally retarded persons employed by sheltered workshops. This population is the subject of a five-year followup study. In conjunction with this investigation, the mental status evaluation for this population is being standardized (Devenny et al., 1991). This evaluation has more detail than that used before by Thase, et al. (1982) and Wisniewski and Hill 1985b).

The current mental status evaluation, is a relatively informal, open ended measure of specific functioning areas, e.g., orientation to person, place, and time; color naming; concentration; motor and amnestic apraxia, and memory; and anomia (Wisniewski and Hill 1985b). It allows for informal observations (e.g., ability to attend to a conversation) as well as structured tasks (e.g., color naming). Each of the questions asked is initially phrased so as to elicit a response in which the individual recalls the information. Items that are failed are repeated in the form of a recognition test in which the individual has a choice of three responses (e.g.

"Is today Sunday, Tuesday, or Friday?"). To assess orientation, three areas are tested: person, place, and time. Items that are recognized but not recalled are given half credit.

Color naming is assessed through the following process. Four colored objects (red, blue, green, and yellow) are presented, and the individual is requested to name the colors. The task is repeated, and a total score of correct naming is obtained. Incorrect substitutions, if closely related (e.g., calling a yellow object "orange") and used consistently, are considered correct. Recognition trials are given by asking the individual to point to the object that has a particular color. Correct recognition responses are given half credit.

To assess concentration, the individual is requested to recite the alphabet and to count forward to 30 and backward from 20. Omissions are scored, and the subject may be prompted within each of the tasks if it appears that he or she is searching for the next item.

To assess for motor and amnestic apraxia, the individual is required first to write his or her name and to print the alphabet. Verbal prompts are allowed if the individual omits letters in the alphabet; written examples are not allowed. He or she is also asked to draw figures from memory. Examples of each figure failed are presented, and the individual is asked to copy each, making his or her drawing the same size as the original. During the presentation, each figure is named. After all figures have been copied, the individual is asked to draw each figure on command (e.g. "draw a circle"). Amnestic apraxia would be suggested by an inability to draw all the figures on command. Scoring is based on qualitative judgments of the individual's ability to draw horizontal, vertical, oblique, and curved lines rather than the figure itself. Mild impairments are indicated by wavy lines, corners failing to meet, hand shakiness, and obvious compensating mechanisms (e.g., making the productions very small); a comparison of the letter writing to any previous examples that may be in the individual's records may also reveal mild impairments.

To assess anomia, the individual is asked to name five objects. Full credit is given if the whole name is provided or if a differentiation can be made between the object (e.g., heel) and the expected confusion (e.g., shoe). Further questioning is allowed (e.g., "Yes, but what do we call this part of the shoe?"). The mental status questions that are primarily for Down syndrome individuals were published previously (Wisniewski and Hill, 1985b).

## DIFFERENTIAL DIAGNOSES

Once dementia has been documented in Down syndrome individuals over the age of 30 years, the etiology must be established. The differential diagnosis is conducted to define whether the dementia is caused by a reversible condition (e.g., reactive depression, endocrine and vitamin deficiencies, uncontrolled seizures) or an irreversible condition [e.g., Alzheimer's disease, subacute sclerotic panencephalitis, Creutzfeldt-Jakob disease, viral central nervous system infection with autoimmune deficiency syndrome (AIDS), adult form of inborn errors of metabolism, Pick's disease, Huntington's chorea]. Alzheimer's disease occurs in more than 50% of all cases of dementia that are associated with old age.

Perhaps the most difficult differential diagnosis is that between Alzheimer's disease and depression. Slightly less difficult is the differential diagnosis between Alzheimer's disease and multi-infarct dementia and between the mixed form of Alzheimer's disease and multi-infarct dementia. One reason for the difficulty in the differential diagnosis is that depression often accompanies dementia of other types. The individual may recognize that he or she can no longer perform as well as before and become depressed by this fact. On the other hand, the decrease in cognitive functioning may be a result of depression. Often, the differential diagnosis is made as a result of the clinical history, time factor, and the response to antidepressants.

Historical information is particularly important in the differential diagnoses between different causes of dementia. Dementia of the Alzheimer's type is relatively progressive and of variable duration (one to 10 years), and there may be periods in which the cognitive level remains at a static level before it continues its progressive decline. Depression and multi-infarct dementia tend to have a relatively rapid onset; severity of cognitive losses exhibited by the affected individuals fluctuates. Depression is usually a reversible condition, whereas multi-infarct dementia may be partially or completely reversible only in some cases and then only with time and appropriate treatment.

It is also important to interview the individual to discover how he or she feels about his or her condition. For instance, depressed people tend to complain about their lives and to exaggerate their failures, whereas individuals with Alzheimer's disease tend to externalize their complaints and to cover up their symptoms. In addition, depressed individuals tend to have a history of depression and respond well to medications. Individuals with Alzheimer's disease typically do not have such a history. Although these people may respond to antidepressants, cognitive symptoms reappear relatively soon. Additional evidence that may help in the differential diagnosis can be obtained from interviewing relatives or significant others who know the person. These people tend to complain more about the memory of depressed persons and about the more generalized disabilities of persons with Alzheimer's disease (Gurland and Toner, 1983). It must be pointed out, however, that progressive memory loss is the most frequent consistent finding in dementia of the Alzheimer's type. Sim and Sussman, (1962) reported that memory loss was the first symptom and dementia has been also confirmed by neurological reports (Coblentz, et al., 1973; Goodman, 1953; Ziegler, 1954).

When attempting a differential diagnosis, one should consider the overlap of symptoms as well as the possibility of mixed etiology (e.g., a diagnosis of Alzheimer's disease does not rule out the possibility of co-occurrence of depression and/or MID). This overlap makes the diagnoses particularly difficult (Liston et al., 1983).

29yrs  30yrs

31yrs                33yrs                39yrs

Fig. 1.  Premature aging of an individual with Down syndrome.

In summary, the diagnosis of Alzheimer's dementia is made on the basis of clinical history, psychological and neurological examinations repeated annually, additional laboratory tests [e.g., electrophysiological: electroencephalogram (EEG), brain stem auditory evoked responses (BAERs), neuroradiological: computer tomography (CT) scan], or magnetic resonance imaging (MRI), or perfusion emission topography (PET) and, if indicated, biochemical studies (e.g., $B_{12}$, folic acid, thyroid, parathyroid levels, SMA 18, CBC) to determine the cause or causes of dementia.

## CLINICAL EVIDENCE OF AGING AND ALZHEIMER'S DISEASE

Accelerated aging is observed in some individuals with Down syndrome. An example of the clinical signs are demonstrated in one 40-year-old Down syndrome person (Fig. 1) who looked older than his chronological age. His clinical history and neuropsychological studies suggested regression, including personality aberrations in the form of apathy, sudden affective changes, deterioration of personal hygiene, loss of vocabulary.

Also, increasingly abnormal neurological signs have been reported in Down syndrome individuals (Dalton et al., 1974; Dalton and Crapper, 1977; Dalton and

Crapper McLachlan, 1984; Loesch-Mizewska, 1968; Owens et al., 1971; Thase et al., 1982, 1984; Wisniewski et al., 1978, 1982; Wisniewski et al., 1987; Wisniewski et al., 1985a,c, and 1987; Wisniewski and Hill, 1985b). Dalton and Crapper McLachlan (1984) reported that recent memory loss, as measured by a delayed matching-to-sample task, consistently was among the first indicators of dementia within an institutionalized population of severely and profoundly mentally retarded individuals with Down syndrome. According to their results, 24% of Down syndrome individuals older than 40 years of age showed signs of memory impairment. Neither the elderly, matched, mental retardation controls nor the younger mental retardation controls, with or without Down syndrome, failed the task.

## AGING AND ALZHEIMER'S DISEASE NEUROPATHOLOGY IN INDIVIDUALS WITH DOWN SYNDROME

The brains of mature and aging persons with Down syndrome appear to show progressive abnormalities in the same neurotransmitter systems as observed in the brains of individuals with Alzheimer's disease in the general population (Mann et al., 1985a,b and 1990; Godridge et al., 1987). These deficits appear to be the largest in the brains that becomes shrunken and also have substantial increased numbers of neuritic senile plaques and neurofibrillary tangles in comparison with age-matched, non-demented controls (Mann et al., 1984). No reports have shown correlations between any neurotransmitter deficits in Down syndrome brains and neuritic senile plaques and neurofibrillary tangle counts, [nor with presence or absence of dementia]. Some investigators have reported small deficits in neurotransmitter levels (Godridge et al., 1987; Yates et al., 1986) in brains of younger subjects without any Alzheimer's disease neuropathology, suggesting that there may be congenital neurotransmitter deficits in Down syndrome (Godridge et al., 1987). Fortunately, investigators are beginning to recognize the need to differentiate between the congenital brain abnormalities of Down syndrome and the degenerative processes of Alzheimer's disease. A brief review of specific neurotransmitter systems in Down syndrome is presented below.

Changes in cholinergic system of Down syndrome brains are quite pronounced, both reduced cell counts and deficits in ChAT concentration in the cortical mantle and the nucleus basalis of Meynert (Shortridge et al., 1985; Casanova et al., 1985; McGeer et al., 1985). Norepinephrine levels and neuronal numbers in the locus coeruleus are also reduced (Yates et al., 1981; Yates et al., 1983; Godridge et al., 1987; Shortridge et al., 1985). Similarly, norepinephrine levels are lower than normal in regions commonly rich in this neurotransmitter (Godridge et al., 1987; Yates et al., 1986) and, correspondingly, cell counts in the raphe nuclei are also reduced (Shortridge et al., 1985). In contrast, dopamine levels are not affected (Yates et al., 1983), although the number of neurons may be reduced in one of the brain stem nuclei (A10 or ventral tegmental area) that sends dopaminergic fibers to the frontal and limbic areas of the cortex (Mann et al., 1987a,b). The nigrostriatal dopamine system does not appear to be affected (Mann et al., 1987a,b). The somatostatin level has also been reported to be lowered in Down syndrome brains (Pierotti et al., 1986). It is agreed that neurotransmitters changes in Down syndrome brains are secondary to congenital and degenerative changes and not a primary defect.

## GENETIC FACTORS IN ALZHEIMER'S DISEASE AND DOWN SYNDROME

Genetic factors appear to play a significant role in both Alzheimer's disease and Down syndrome, but the mode of transmission—autosomal-dominant or polygenic—has not been established. The data available on the cumulative incidence of Alzheimer's disease among first degree relatives vary so widely that, for all practical purposes, we do not yet know the magnitude of the risk of inheriting Alzheimer's disease (Larsson et al., 1963; Heston et al., 1981, 1991; Heyman et al., 1983; Breitner et al., 1984; Martin et al., 1988; Schupf et al., 1990).

There is evidence for a familial link between Alzheimer's disease and Down syndrome: families with members who have died with Alzheimer's disease have shown a higher rate of births with Down syndrome (3.5/1000) than families with no history of Alzheimer's disease (1/1000) (Heston et al., 1981). That early-onset Alzheimer's disease and Down syndrome cluster together was recently confirmed by a study of the frequency of Alzheimer's disease in families with Down syndrome: the frequency of early-onset, but not late-onset, Alzheimer's disease was significantly increased in these families (Yatham et al., 1988).

There is no evidence for consistent chromosomal abnormalities in persons with Alzheimer's disease (Glenner, 1988). On the other hand, the chromosomal basis of Down syndrome is well known: 96% of Down syndrome individuals have chromosome 21 in triplicate; a small number of Down syndrome individuals with the characteristic features of Down syndrome indicate that the critical portion is 21q22. How this imbalance produces Down syndrome with its congenitally abnormal brain and Alzheimer's disease neuropathology later in life is not known. Several gene loci on chromosome 21 have been identified in Down syndrome through the presence of increased amounts of gene products; among them are Cu and Zn superoxide dismutase (Epstein, 1986).

The past decade has been particularly exciting in the area of the molecular genetics of Alzheimer's disease and Down syndrome. Several studies have mapped a gene that codes for β-amyloid to human chromosome 21 (Robakis et al., 1987; Goldgaber et al., 1987; Kang et al., 1987; Tanzi et al., 1987). This gene was seen to be of direct relevance to the production of excessive amyloid β-plaques and amyloid angeopathy in Down syndrome and Alzheimer's disease and was seen as providing a genetic link between Alzheimer's disease and Down syndrome. This perception was further strengthened by reports suggesting that a gene responsible for familial Alzheimer's disease also maps to chromosome 21 and that the β-amyloid gene on chromosome 21 (St. George-Hyslop et al., 1987) is duplicated in Alzheimer's disease (Delebar et al., 1987; Schweber, 1985).

Subsequent reports, however, have not replicated either finding linking chromosome 21 to Alzheimer's disease. A study of Alzheimer's disease families failed to establish linkage to chromosome 21 for both early-onset and late-onset Alzheimer's disease, including the region where early-onset Alzheimer's disease had been localized (St. George-Hyslop et al., 1987; Schellenberg et al., 1988). Other studies failed to find three copies of the β-amyloid gene in Alzheimer's disease patients (Warren et al., 1987; Furuya et al., 1988; Podlisny et al., 1987). Correspondingly, no elevation in β-amyloid gene dose has been found in Alzheimer's disease, whereas the expected 1.5 times gene dosage has been observed in Down syndrome (Furuya et al., 1988; Podlisny et al., 1987). However, recently, an English group of investigators reported a point mutation in β-protein in two of 17 early-onset families studied. Because the mutation occurs so rarely and was noted only in early-onset Alzheimer's disease, which is so rare that its prevalence has not even been quantified, it cannot be the cause of Alzheimer's disease in all cases. In summary, there is probably no such thing as a simple Alzheimer gene. Alzheimer's disease is likely to be the result of a defect in a gene or genes influenced by environmental factors. Also, of practical importance is a recent report that the brain macrophages/phagocytes (microglia and pericytes) are the cells that make the β-amyloid fibrils. Now these cells can become the target for Alzheimer's disease therapy (Wisniewski et al., 1990). Therefore, future genetic and biochemical studies are needed to better characterize these suggestions.

## ACKNOWLEDGMENTS

The authors wish to express their appreciation to Mr. Lawrence Black for bibliographical assistance, Ms. Maureen Stoddard Marlow for editorial assistance, and Ms. Madeline Tinney for secretarial assistance. This work was supported by NIH grant NICHD 5 PO1 HD22634.

## REFERENCES

Allison RS (1961). Chronic amnesic syndromes in the elderly. Proc R Soc Med 54:961-965.
American Psychiatric Association (1987). Diagnostic and statistical manual of mental disorders: DSM-III-R. Washington, DC, 3rd ed. rev., p 567.
Baird PA, Sadovnick AD (1987). Life expectancy in Down syndrome. J Pediatr 110:849-854.
Bilash I, Zubek JP (1960). The effects of age on factorially "pure" mental abilities. J Gerontol 15:175-182.
Birren JE (1963). Research on the psychologic aspects of aging. Geriatrics 18:393-403.
Breitner JCS, Folstein MF (1984). Familial Alzheimer Dementia: a prevalent disorder with specific clinical features. Psychol Med 14:63-80.
Burger PC, Vogel FS (1973). The developmental of the pathologic changes of Alzheimer's disease and senile dementia in patients with Down's syndrome. Am J Pathol 73:457-476.
Casanova MF, Walker LC, Whitehouse PJ, Price DL (1985). Abnormalities of the nucleus basalis in Down's syndrome. Ann Neurol 18:310-313.

Clark JW (1960). The aging dimension: a factorial analysis of individual differences with age on psychological and physiological measurements. J Gerontol 15:183-187.

Coblentz JM, Mattis S, Zingesser LH, Kasoff SS, Wisniewski HM, Kutzman R (1973). Presenile dementia. Clinical aspects and evaluation of cerebrospinal fluid dynamics. Arch Neurol 29:299-308.

Coyle JT, Oster-Granite ML, Gearhart JD (1986). The neurobiologic consequences of Down syndrome. Brain Res Bull 16:773-787.

Crome L, Stern J (1972). Pathology of Mental Retardation. 2d ed. Edinburgh, Churchill Livingstone.

Dalton AJ, Crapper DR, Schlotterer GR (1974). Alzheimer's disease in Down's syndrome: visual retention deficits. Cortex 10:366-377.

Dalton AJ, Crapper DR (1977). Down's syndrome and aging of the brain. In: P. Mittler (ed). Research to practice in mental retardation: Biomedical Aspects. Baltimore, University Park Press, 3:391-400.

Dalton AJ, Crapper McLachlan DR (1984). Incidence of memory deterioration in aging persons with Down's syndrome. In: Berg JM (ed). Perspectives and Progress in Mental Retardation, 2 (Baltimore, MD, University Park Press), pp 55-62.

Delabar JM, Goldgaber D, Lamour Y, Nicole A, Huret JL, de Grouchy J, Brown P, Gajdusek DC, Sinet PM (1987). Beta amyloid gene triplication in Alzheimer's disease and karyotypically normal Down syndrome. Science 235:1390-1392.

Devenny DA, Hill AL, Patxot O, Silverman WP, Wisniewski KE (1991). Ageing in higher functioning adults with Down syndrome: an interim report in a longitudinal study. J Ment Defic Res (in press).

Epstein CJ (1986). Trisomy 21 and the nervous system: from cause to cure. In: Epstein CJ (ed): The Neurobiology of Down Syndrome. New York, Raven, p 1.

Fenner ME, Hewitt KE, Torpy DM (1987). Down's syndrome: intellectual and behavioural functioning during adulthood. J Ment Defic Res 31:241-249.

Fraser J, Mitchell A (1876). Kalmuc idiocy: report of a case with autopsy with notes on 62 cases. J Ment Sci 22:161.

Furuya H, Sasaki H, Goto I, Wong CW, Glenner GG, Sakaki Y (1988). Amyloid beta-protein gene duplication is not common in Alzheimer's disease: analysis by polymorphic restriction fragments. Biochem Biophys Res Commun 150:75-81.

Glenner GG (1988). Alzheimer's disease: its proteins and genes. Cell 52:307-308.

Goate A, Chartier-Harlin MC, Mullan M, et al. (1991). Segregation of a missense mutation in the amyloid precursor protein gene with familial Alzheimer's disease. Nature 349:704-706.

Godridge H, Reynolds GP, Czudek C, Calcutt NA, Benton M (1987). Alzheimer-like neurotransmitter deficits in adult Down's syndrome brain tissue. J Neurol Neurosurg Psychiatry 50:775-778.

Goldgaber D, Lerman MI, McBride OW, Saffiotti U, Gajdnsek DC (1987). Characterization and chromosomal localization of a cDNA encoding brain amyloid of Alzheimer's disease. Science 235:877-880.

Goldstein G, Shelly CH (1975). Similarities and differences between psychological deficit in aging and brain damage. J Gerontol 30:448-455.

Goodman L (1953). Alzheimer's disease: clinico-pathologic analysis of 23 cases with a theory on pathogenesis. J Nerv Ment Dis 118:97-130.

Gurland B, Toner J (1983). Differentiating dementia from nondementing conditions. Adv Neurol 38:1-17.

Henderson AS (1986). The epidemiology of Alzheimer's disease. Br Med Bull 42:3-10.

Heston LL, Mastri AR, Anderson VE, White J (1981). Dementia of the Alzheimer type. Clinical genetics, natural history and associated conditions. Arch Gen Psychiatry 38:1085-1091.

Heston LL, Orr HT, Rich SS, White JA (1991). Linkage of an Alzheimer disease susceptibility locus to markers on human chromosome 21. Am J Med Genet (in press).

Heyman A, Wilkinson WE, Hurwitz BJ, Schmechel D, Sigmon AH, Weinberg T, Nelms MJ, Swift M (1983). Alzheimer's disease: genetic aspects and associated clinical disorders. Ann Neurol 14:507-515.

Jarvik LF (1975). Thoughts on the psychobiology of aging. Am Psychol 30:576-583.

Kang J, Lemaire HG, Unterbeck A, Salbaum JM, Masters CL, Grzechik KH, Multharp G, Beyreuther K, Müller-Hill B (1987). The precursor of Alzheimer's disease amyloid A4 protein resembles a cell-surface receptor. Nature 325:733-736.

Katzman R (1981). Early detection of senile dementia. Hosp Pract [off] 16(6):61-76.

Lai F, Williams R (1987). Alzheimer's dementia in Down's syndrome. (Abstract) Neurology 37:332.

Larsson T, Sjogren T, Jacobson G (1963). Senile dementia. A clinical, sociomedical and genetic study. Acta Psychiatr Scand Suppl 167:1-259.

Lezak MD (1976). Neuropsychological assessment. New York: Oxford University Press.

Liston EH, La Rue A (1983). Clinical differentiation of primary degenerative and multi-infarct dementia: a critical review of the evidence. Part I: Clinical studies. Biol Psychiatry 18:1451-1465.

Loesch-Mdzewska D (1968). Some aspects of the neurology of Down's syndrome. J Ment Defic Res 12:237-246.

Lott IT (1982). Down's syndrome, aging, and Alzheimer's disease: a clinical review. Ann NY Acad Sci 396:15-27.

Malamud N (1972). Neuropathology of organic brain syndromes associated with aging. In: Gaitz CM (ed). Aging and the Brain, New York, Plenum, p 63.

Mann DMA, Yates PO, Marcyniuk B (1984). Alzheimer's presenile dementia, senile dementia of Alzheimer type and Down's syndrome in middle age form an age related continuum of pathological changes. Neuropathol Appl Neurobiol 10:185-207.

Mann DMA, Yates PO, Marcyniuk B (1985a). Some morphometric observations on the cerebral cortex and hippocampus in presenile Alzheimer's disease, senile dementia of Alzheimer type and Down's syndrome in middle age. J Neurol Sci 69:139-159.

Mann DMA, Yates PO, Maracyniuk B, Ravindra CR (1985b). Pathological evidence for neurotransmitter deficits in Down's syndrome of middle age. J Ment Defic Res 29:125-135.

Mann DMA, Yates PO, Marcyniuk B, Ravindra CR (1987a). Loss of neurons from cortical and subcortical areas in Down's syndrome patients at middle age. Quantitative comparisons with younger Down's patients and patients with Alzheimer's disease. J Neurol Sci 80:79-89.

Mann DMA, Yates PO, Marcyniuk B (1987b). Dopaminergic neurotransmitter systems in Alzheimer's disease and in Down's syndrome at middle age. J Neurol Neurosurg Psychiatr 50:341-344.

Mann DMA, Royston MC, Ravindra CR (1990). Some morphometric observations on the brains of patients with Down's syndrome: their relationship to age and dementia. J Neurol Sci 99:153-164.

Marin-Padilla M (1976). Pyramidal cell abnormalities in the motor cortex in a child with Down's syndrome. A Golgi study. J Comp Neurol 167:63-81.

Martin RL, Gerteis G, Gabrielli WF Jr (1988). A family-genetic study of dementia of Alzheimer type. Arch Gen Psychiatry 45:894-900.

McGeer EG, Norman M, Boyes B, O'Kusky J, Suzuki J, McGeer PL (1985). Acetylcholine and aromatic amine systems in postmortem brain of an infant with Down's syndrome. Exp Neurol 87:557-570.

Mortimer JA (1983). Alzheimer's disease and senile dementia: prevalence and incidence. In: Reisberg B (ed). Alzheimer's Disease. New York, Free Press, p 141.

Oliver C, Holland AJ (1986). Down's syndrome and Alzheimer's disease: a review. Psychol Med 16:307-322.

Opitz JM, Gilbert-Barness EF (1990). Reflections on the pathogenesis of Down syndrome. Am Med Genet, Suppl 7, pp 38-51.

Owens D, Dawson JC, Losin S (1971). Alzheimer's disease in Down's syndrome. Am J Ment Defic 75:606-612.

Pierotti AR, Harmar AJ, Simpson J, Yates CM (1986). High-molecular-weight forms of somatostatin are reduced in Alzheimer's disease and Down's syndrome. Neurosci Lett 63:141-146.

Podlisny MB, Lee G, Selkoe DJ (1987). Gene dosage of the amyloid beta precursor protein in Alzheimer's disease. Science 238:669-671.

Purpura DP (1974). Dendritic spine "dysgenesis" and mental retardation. Science 186:1126-1128.

Rabe A, Wisniewski KE, Schupf N, Wisniewski HM (1990). Relationship of Down's syndrome to Alzheimer's disease. In: Deutsch SI, Weizman A, Weizman R (eds). Application of Basic Neuroscience to Child Psychiatry. Plenum Publ. Corp., pp 325-340.

Reitan RM (1967). Psychologic changes associated with aging and with cerebral damage. Mayo Clin Proc 42:653-673.

Robakis NK, Wisniewski HM, Jenkins EC, Devine-Gage EA, Houck GE, Yao XL, Ramakrishna N, Wolfe G, Silverman WP, Brown WT (1987). Chromosome 21q21 sublocalisation of gene encoding beta-amyloid peptide in cerebral vessels and neuritic (senile) plaques of people with Alzheimer's disease and Down syndrome. [Letter] Lancet I:384-385.

Ross MH, Galaburda AM, Kemper TL (1984). Down's syndrome. Is there a decreased population of neurons? Neurology 34:909-916.

Schellenberg GD, Bird TD, Wijsman EM, Moore DK, Boehnke M, Bryant EM, Lampe TH, Nochlin D,

Schmidt-Sidor B, Wisniewski KE, Shepard TH, Sersen EA (1990). Brain growth and maturation in Down's syndrome subjects 15 to 22 weeks of gestational age and birth to 60 months. Clin Neuropathol 9:181-190.

Schupf N, Zigman WB, Silverman WP, Rabe A, Wisniewski HM (1990). Genetic epidemiology of Alzheimer's disease. In: Battistin L, Gerstenbrand F (eds). Aging Brain and Dementia: New Trends in Diagnosis and Therapy. New York, Wiley-Liss, pp 57-78.

Schweber M (1985). A possible unitary genetic hypothesis for Alzheimer's disease and Down syndrome. Ann NY Acad Sci 450:223-238.

Scott BS, Petit TL, Becker LE, Edwards BAV (1981). Abnormal electric membrane properties of Down's syndrome DRG neurons in cell culture. Brain Res 254:257-270.

Scott BS, Becker LE, Petit TL (1983). Neurobiology of Down's syndrome. Prog Neurobiol 21:199-237.

Shortridge BA, Vogel FS, Burger PC (1985). Topographic relationship between neurofibrillary change and acetylcholinesterase rich neurons in the upper brain stem of patients with senile dementia of the Alzheimer's type and Down's syndrome. Clin Neuropathol 4:227-237.

Sim M, Sussman I (1962). Alzheimer's disease: its natural history and differential diagnosis. J Nerv Ment Dis 135:489-499.

St. George-Hyslop PH, Tanzi RE, Polinsky RJ, Haines JL, Nee L, Watkins PC, Myers RH, Feldman RG, Pollen D, Drachman D, Growdon J, Bruni A, Foncin JF, Salmon D, Frommelt P, Amaducci L, Sorbi S, Piacentini S, Stewart GD, Hobbs WJ, Conneally PM, Gusella JF (1987). The genetic defect causing familial Alzheimer's disease maps on chromosome 21. Science 235:885-890.

Suetsugu M, Mehraein P (1980). Spine distribution along the apical dendrites of the pyramidal neurons in Down's syndrome. A quantitative golgi study. Acta Neuropathol (Berl) 50:207-210.

Sumi SM, Deeb SS, Beyreuther K, Martin GM (1988). Absence of linkage of chromosome 21q21 markers to familial Alzheimer's disease. Science 241:1507-1510.

Sylvester PE (1983). The hippocampus in Down's syndrome. J Ment Defic Res 27:227-236.

Sylvester PE (1986). The anterior commissure in Down's syndrome. J Ment Defic Res 30:19-26.

Takashima S, Becker LE, Armstrong DL, Chan F (1983). Abnormal neuronal development in the visual cortex of the human fetus and infant with Down's syndrome: a quantitative and qualitative Golgi study. Brain Res 225:1-21.

Tanzi RE, Gusella JF, Watkins PC, Bruns GA, St George-Hyslop P, Van Keuren ML, Patterson D, Pagan S, Kurnit DM, Neve RL (1987). Amyloid beta protein gene: cDNA, mRNA distribution, and genetic linkage near the Alzheimer locus. Science 235:880-884.

Thase ME, Liss L, Smeltzer D, Maloon J (1982). Clinical evaluation of dementia in Down's syndrome: a preliminary report. J Ment Defic Res 26:239-244.

Thase ME, Tigner R, Smeltzer DJ, Liss L (1984). Age-related neuropsychological deficits in Down's syndrome. Biol Psychiatry 19:571-585.

Warren AC, Robakis NK, Ramakrishna N, Koo EH, Ross CA, Robb AS, Folstein MF, Price DL, Antonarakis SE (1987). Beta-amyloid gene is not present in three copies in autopsy-validated Alzheimer's disease. Genomics 1:307-312.

Whalley LJ (1982). The dementia of Down's syndrome and its relevance to aetiological studies of Alzheimer's disease. Ann NY Acad Sci 396:39-53.

Williams M (1970). Geriatric patients. In: P. Mittler (ed). The psychological assessment of mental and physical handicaps. London: Methuen, pp 319-339.

Wisniewski HM, Rabe A, Wisniewski KE (1987). Neuropathology and dementia in people with Down's syndrome. In: Davies P, Finch C (eds.): Banbury Report 27: Molecular Neuropathology of Aging. Cold Spring Harbor, New York: Cold Spring Harbor Laboratory, p 399.

Wisniewski K, Howe J, Williams DG, Wisniewski HM (1978). Precocious aging and dementia in patients with Down's syndrome. Biol Psychiatry 13:619-627.

Wisniewski KE, French JH, Rosen JF Kozlowski PB, Tenner M, Wisniewski HM (1982). Basal ganglia calcification (BGC) in Down's syndrome (DS)—another manifestation of premature aging. Ann NY Acad Sci 396:179-189.

Wisniewski KE, Laure-Kamionowska M, Wisniewski HM (1984). Evidence of arrest of neurogenesis and synaptogenesis in brains of patients with Down's syndrome. N Engl J Med 311:1187-188 (Letter).

Wisniewski KE, Dalton AJ, Crapper McLachlan DR, Wen GY, Wisniewski HM (1985a). Alzheimer's disease in Down's syndrome: clinicopathologic studies. Neurology 35:957-961.

Wisniewski K, Hill AL (1985b). Clinical aspects of dementia in mental retardation and developmental disabilities, in Janicki MP, Wisniewski HM (eds): Aging and Developmental Disabilities. Issues and Approaches, Baltimore, Brookes, p 195.

Wisniewski KE, Wisniewski HM, Wen GY (1985c). Occurrence of neuropathological changes and dementia of Alzheimer's disease in Down's syndrome. Ann Neurol 17:278-282.

Wisniewski KE, Laure-Kamionowska M, Connell F, Wen GY (1986). Neuronal density and synaptogenesis in the postnatal stage of brain maturation in Down syndrome. In: Epstein CJ (ed): The Neurobiology of Down Syndrome. New York, Raven, p 29.

Wisniewski K, Schmidt-Sidor B (1986). Myelination in Down's syndrome brains (pre- and postnatal maturation) and some clinical pathological correlations. (Abstract) Ann Neurol 20:429-430.

Wisniewski KE, Miezejeski CM, Hill AL (1988). Neurological and psychological status of individuals with Down syndrome. In: Nadel L (ed). The Psychobiology of Down Syndrome. Cambridge, Mass, The MIT Press, pp 315-343.

Wisniewski KE (1990). Down syndrome children often have brain with maturation delay, retardation of growth, and cortical dysgenesis. Am J Med Genet Suppl 7:274-281.

Yates CM, Ritchie IM, Simpson J, Maloney AF, Gordon A (1981). Noradrenaline in Alzheimer-type dementia and Down syndrome. [Letter] Lancet 2:39-40.

Yates CM, Simpson J, Gordon A, Maloney AF, Allison Y, Ritchie IM, Urquhart A (1983). Catecholamines and cholinergic enzymes in pre-senile and senile Alzheimer-type dementia and Down's syndrome. Brain Res 280:119-126.

Yates CM, Simpson J, Gordon A (1986). Regional brain 5-hydroxytryptamine levels are reduced in senile Down's syndrome as in Alzheimer's disease. Neurosci Lett 65:189-192.

Yatham LN, McHale PA, Kinsella A (1988). Down's syndrome and its association with Alzheimer's disease. Acta Psychiatr Scand 77:38-41.

Ziegler DK (1954). Cerebral atrophy in psychiatric patients. Am J Psychiatry 111:454-458.

Zigman WB, Schupf N, Lubin RA, Silverman WP (1987). Premature regression of adults with Down syndrome. Am J Ment Defic 92:161-168.

# Index

Achenbach Child Behavior Checklist, behavioral characteristics assessed via, 58
Achondroplasia, otitis media associated with, 88
Acidosis, due to untreated sleep apnea, 132
Acrocephalosyndactyly, otitis media associated with, 88
Acute corneal hydrops, incidence of, 147, 152
Acute lymphoblastic leukemia (ALL), increased susceptibility to, 94-96
Acute megakaryoblastic leukemia (AMKL), increased susceptibility to, 93-99
Acute non-lymphoblastic leukemia (ANLL), increased susceptibility to, 93-94, 96
*AD1* gene, on chromosome 21, 9
Adaptive functioning, decline in, 104-105
Adductor release, coupled with first metatarsal osteotomy, for hallux valgus, 114
Adenoidectomy, for obstructive sleep apnea, 57, 131-132
Adenoids, in alveolar hypoventilation, 65
Adjustment reactions, incidence of, 57
Adolescence, tasks confronted in, 158-160
Afebrile seizures, incidence of, 105
Affectionate behavior, strong sense of, 106
Affective disorders, incidence of, 106
AFP. *See* Alpha-fetoprotein
Aganglionic megacolon, incidence of, 55
Aggressive behavior, in school age children, xvi
Aging
    Alzheimer's disease and, xii, 167-178
    health maintenance and, xvi
    motor competence decline and, 104
Air tympanometry examination, of middle ear, 130
Airway infections, increased susceptibility to, 83

Albers–Sconberg disease, otitis media associated with, 88
ALL. *See* Acute lymphoblastic leukemia
Allen cards, in assessment of visual function, 153
Alopecia, incidence of, 87
Alpha-fetoprotein (AFP), as biochemical screening parameter for Down syndrome, 14, 17-19
*ALS* gene, on chromosome 21, 8-9
Alveolar hypoplasia, incidence of, 64
Alveolar hypoventilation
    factors predisposing to, 65
    incidence of, 64-65
    mechanisms of action of, 65
Alzheimer's disease
    aging and, 167-178
    amyloid precursor protein and, 8
    clinical evidence of, 175-176
    genetic factors in, 177-178
    neuropathology of, 104-105, 176-177
    screening for, xvi
Amblyopia, incidence of, 150-152, 154
American Association on Mental Deficiency's Adaptive Behavior Scales, in assessment of social and behavioral functioning, 172
AMKL. *See* Acute megakaryoblastic leukemia
Amnestic apraxia, assessment for, 173
Amnestic indifference, characterization of, 171
Amniocentesis, Down syndrome assessed via, 13, 19
ß-Amyloid gene, in Alzheimer's disease and Down syndrome, 177-178
Aneuploidy
    deleterious effects of, 10
    index for in prenatal diagnosis, 18
    phenotypic studies of, 104

Animal models, in study of Down syndrome, x
Anisometropia, incidence of, 149, 151, 153
ANLL. *See* Acute non-lymphoblastic leukemia
Annular pancreas, incidence of, 55
Anomia, assessment for, 173
Anterior commissure, reduction in size of, 169
Anterior rugae, prominent, 135, 140
Antibiotics, against middle ear effusion, 129
Anticonvulsive medications, against seizures, 56, 105
Antimicrosomal antibodies, in thyroid dysfunction, 73-74
Antithyroglobulin antibodies, in thyroid dysfunction, 73-74
Anus, imperforated, occurrence of, 55
Anxiety disorders, incidence of, 58
Aortic stenosis, incidence of, 61-62
Apert syndrome, otitis media associated with, 88
Apnea, obstructive, 57, 64-66, 127, 130-132
*APP* gene, on chromosome 21, 7-9
Apraxia, assessment for, 173
Arch supports
    flexible, 113
    response to, 114
Arginine, effect of on plasma growth hormone concentrations, 79-80
Astigmatism, incidence of, 147, 149
Athletic activities
    patellofemoral instability and, 116-117
    spinal problems and, 123, 125
Atlanto-axial instability
    incidence of, xv, 57, 65, 105-107
    spinal problems and, 121, 124
Atrial septal defect, incidence of, 55, 62-64
Atrioventricular canal defect, incidence of, 55, 61-62, 67
Attention demands, excessive, 58
Audiogram
    of middle and inner ear, 49
    in preschoolers, xvi, 3
    in school age children, xvi, 5
Auditory canal, external, diameter of, 127
Auditory memory, deficits in, 104, 107
Auditory trainer, amplification through use of, 50
Australia antigen, discovery of, 84

Autism, infantile, Down syndrome associated with, 107
Autoimmune disorders, incidence of, 87
Automobile seat, high-backed, for protection of cervical spine, 123, 125
Autoradiogram, for quantitative analysis of gene copy number in patients with partial trisomy 21, 6
Autosomal trisomies, human, nondisjunction in, 23-31

Baby bottle syndrome, incidence of, 141
Bacterial antigens, defective cellular responses to, 86
Bacterial tracheitis, Down syndrome association with, 87
BAERs. *See* Brain stem auditory evoked responses
Baltimore Washington Infant Study, congenital heart disease assessed via, 61-62
Basal ganglia, calcification of, 103, 169
Basal osteotomy, of first metatarsal, for bunions, 115
Basal spiral tract, progressive osteogenesis along outflow pathway of, 128
Basilar osteotomies, of first metatarsal, 113
Basophil granules, in megakaryoblasts, 98-99
*BCEI* gene, on chromosome 21, 8-9
B-cells, reduction in number of, 86
Bed wetting, incidence of, 66
Behavioral disorders, incidence of, 57-58, 163-164
Big toe, "hitchhiking", 112
Biochemical screening parameters
    alpha-fetoprotein, 14, 17-19
    human chorionic gonadotropin, 14-19
    multiparameter approaches, 17
    pregnancy specific glycoprotein, 16-17, 19
    unconjugated estriol, 16-17, 19
Blepharitis, incidence of, 56, 147-148, 154
Blindness, incidence of, 147, 153
Bone marrow, fibrosis of, in leukemia, 96
Bone tympanometry examination, of middle ear, 130
Braces
    for bilateral instability of tibio-femoral joint, 117
    for scoliosis, 121
Brachycephaly, on chromosome 21 phenotypic map of Down syndrome, 7

Brain
abnormalities of, xi, 168-170
functional studies in, 104
structural studies of, 103-104
Brain stem auditory evoked responses
(BAERs)
in Alzheimer's dementia, 175
in neonatal and early infancy, xv
Brain stem nuclei, neuronal numbers in, 176
Brückner test, for eye evaluation, 152-153
Brushfield spots, incidence of, 147-148
Bruxism, incidence of, 135, 140, 142
Bunions
complex, 124
early, 114
in preschoolers, xv
surgical correction of, 114-115

Cancer, increased susceptibility to, xvi, 100
Capital femoral epiphysis, slipped, 119-120,
124
Capsular tissues, about major joints, 111
Capsuloplasty
for bunions, 115
for dislocated hip, 118
for hallux valgus, 114
Cardiac pathophysiology, in Down
syndrome, 63
Cardiorespiratory problems, incidence of, xi,
61-68; see also specific problems
Cataracts, incidence of, xv, 55, 147, 152-154
Cattell Infant Intelligence Scales, in
assessment of functioning levels, 171
Causal constructs, in accounting for
language deficits, 40-41, 44
CBS gene, on chromosome 21, 8-9
CD18 gene, on chromosome 21, 8-9
Cell adhesiveness, abnormal, 62
"Cell therapy", in attempted modification of
Down syndrome phenotype, 88-89
Cellular immunity, defective, in Down
syndrome, 85, 90, 100
Cerebral cortex, reduced number of neurons
in, 169
Cerebral granular layers II and IV, defect in,
103
Cervical spine, orthopedic problems in, xvi,
121-125
Chiari pelvic osteotomy, for dislocated hip,
119
Cholesteatoma, incidence of, 129

Choline acetyltransferase, concentrations of,
in cortical mantle, 176
Cholinergic neurotransmitter systems,
disturbances of, 104, 176
Chorionic villi sampling, Down syndrome
assessed via, 13, 19
Chromosomal karyotyping, in neonatal and
early infancy, xv
Chromosome 21, in Down syndrome
genes for, 6-9
genotype-phenotype correlations and, 4-5
history of, 3-4
mental retardation genes and, 8, 10
molecular structure of, 5-6
phenotypic map of, 6-8
physical mapping of, x, 1
third in genome, xi
Circulatory disturbances, incidence of, 62-63
Cleft palate, otitis media associated with, 88
Clonidine, growth hormone release
stimulated by, 80
Coarctation, incidence of, 61-62
Cognitive deficits
auditory memory and, 104
characterization of, 1, 10, 37-38
visual sequential memory and, 104
COL6A1/2 genes, on chromosome 21, 8-9
Communication skills, in preschoolers, xvi
Community living, in Down syndrome, xii,
157-165
Computed tomography (CT), Alzheimer's
dementia assessed via, 175
Computer-based remedial programs,
tailoring of to specific needs of
children, 38
Concanavalin A, decreased T-cell
proliferative response of to, 86
Concentration difficulties, incidence of, 58
Conduct disorders, incidence of, 57
Congestive heart failure, incidence of, 62
Connective tissue, poor, 111
Contraceptives, choices of, 77, 160-161
Cranial sonograms, thalamic echogenicity
assessed via, 103
Craniofacial abnormalities, otitis media
associated with, 87-88
Creutzfeldt–Jakob disease, dementia
associated with, 173
Crouzon disease, otitis media associated
with, 88
CRYA1 gene, on chromosome 21, 8-9
CT. See Computed tomography

Cyanosis, incidence of, xv, 63
Cystic hygroma, ultrasound assessment of, 19
Cytogenetic studies, of human autosomal trisomies, 25
Cytokines, depressed production of, 86

D17S33 DNA, on chromosome 21, 6
D21S series DNA, on chromosome 21, 6, 9, 27
Daily living skills, loss of, in adults, xvi
Dalton and Crapper Test, in assessment of dementia, 172
Daytime somnolence, due to untreated sleep apnea, 132
Deinstitutionalization, as social change, ix
Dementia
    assessment of, 170-171
    definition of, 170
    differential diagnoses of, 173-175
    multi-infarct, 174
Dendritic spines, dysgenesis of, 169-170
Dental considerations, in Down syndrome, xi, xv, xvi, 135-145; see also specific considerations
Depression
    dementia associated with, 173-174
    incidence of, xvi, 57-58, 106, 163
Dermal folliculitis, increased susceptibility to, 83
Dermatoglyphics, on chromosome 21 phenotypic map of Down syndrome, 7
Developmental evaluation, of preschoolers, xvi
Developmental retardation, Down syndrome associated with, x
Diabetes mellitus
    insulin-dependent, 81
    human chorionic gonadotropin production in, 16
    screening for, xvi
Dietary counseling, for school age children and young adults, xvi
Disruptive behaviors, incidence of, 58
Distal osteotomies, of first metatarsal, 113
DNA markers, in testing "relaxed selection" hypothesis, 29
DNA polymorphisms, Down syndrome associated with, xi
DNA studies, of human autosomal trisomies, 25-26

Dopaminergic neurotransmitter systems, disturbances of, 104, 176-177
DPT vaccine, need for, 89-90
Duodenal atresia
    incidence of, 55
    ultrasound assessment of, 19
Dyserythropoiesis, morphologic features of, in leukemia, 98

Ear canal, external, variations in shape and size of, 127
Ear infections, increased susceptibility to, 49, 56, 83, 90
Early intervention program, enrollment in, xv, xvi
Eating behavior, in preschoolers, xvi
Echocardiogram, of congenital heart disease, xv, 55, 66
Educational strategies, in Down syndrome, 53, 58-59; see also specific strategies
EEG. See Electroencephalogram
"E"-game, in assessment of visual function, 153
Eisenmenger syndrome, incidence of, 63-64, 68
EKG. See Electrocardiogram
Electrocardiogram (EKG), in neonatal and early infancy, xv
Electroencephalogram (EEG), Alzheimer's dementia assessed via, 175
Employment placement, of people with Down syndrome, 59
Enamel, defects of, 135
Endocrine deficiencies
    dementia associated with, 173
    incidence of, xi, 71-81
EPM1 gene, on chromosome 21, 8-9
ERG gene, on chromosome 21, 8-9
Erythroleukemia, increased susceptibility to, 96-97
Esotropia, incidence of, 149-150
Estriol, unconjugated (uE3), as biochemical screening parameter for Down syndrome, 16-17, 19
Ethics, of genetic amniocentesis, 13
ETS 2 gene, on chromosome 21, 7-9
Eustachian tubes
    congenital anomalies of, 56, 87, 90, 128
    variations in shape and size of, 127
Exercise program
    for adults, xvi

plasma growth hormone concentrations and, 79
for school age children, xvi
Extracellular matrix, migration of, 62
Eye disorders. *See* Ocular disorders

Facies, on chromosome 21 phenotypic map of Down syndrome, 7
Family history, of Down syndrome, 14
Fascial layers, thin, 111
"Fast mapping", as vocabulary acquisition strategy, 48
Febrile seizures, incidence of, 105
Feeding evaluation, of preschoolers, xvi
Feet, orthopedic problems of, xv, 57, 112-116, 124
Femur length, shortened, ultrasound assessment of, 19
Fertility, and reproduction issues, 74, 77-78
Fetuses, delayed brain development in, 169
Finger, incurved fifth, on chromosome 21 phenotypic map of Down syndrome, 7
First metatarsal, S-shaped, 113-114
First trimester screening, for Down syndrome, 18
Fixation targets, in assessment of visual function, 153
Fluoride supplementation, against tooth decay, xv, 143, 145
Follicle stimulating hormone (FSH), in gonadal function, 75-76
FSH. *See* Follicle stimulating hormone
Fungal infections, increased susceptibility to, 83

Gait changes, in adults, xvi
*GART* gene, on chromosome 21, 8
Gastrointestinal tract, congenital anomalies of, xv, 55
G-banded chromosome 21 heteromorphisms, analysis of, 25
Genes, chromosome 21, 6-9
Genetic factors, in Alzheimer's disease and Down syndrome, 177-178
Genetic maps, based on female meioses leading to trisomy 21, 32
Genetic recombination, non-disjunction associated with, 30-31
Genioglossus muscle, hypotonicity of, 131
Genotype-phenotype correlations, current status of, 4-5
Genu valgum, severe, 115

GHRF. *See* Growth hormone releasing factor
Gingivitis, incidence of, 142, 144
Glossoptosis
alveolar hypoventilation and, 65
otitis media and, 88
Goldenhar syndrome, otitis media associated with, 88
Gonadal function
in Down syndrome subjects, 74-78
otitis media and, 88
Granulocyte progenitor cells, in blood, abnormally low levels of, 99
Granulocytes, circulating, abnormalities of, 99
Great arteries, transposition of, 61-62
Growth hormone (GH), levels of, 79-81
Growth hormone releasing factor (GHRF), decreased release of, 80
Growth patterns, in Down syndrome, 74, 78-81
Gut atresia, on chromosome 21 phenotypic map of Down syndrome, 7

Hallux varus, incidence of, 112-113
hCG. *See* Human chorionic gonadotropin
Head infections, increased susceptibility to, 83
Health care planning, for individual with Down syndrome, xii
Hearing aids, amplification through use of, 50
Hearing deficits, incidence of, xv, 49-50, 56, 84, 127-130
Heart
on chromosome 21 phenotypic map of Down syndrome, 7
disease
in alveolar hypoventilation, 65
incidence of, 10, 61-68
in neonatal and early infancy, xv, 53, 55
management of, 66-67
congenital, surgery for, 67
failure, 55
irregular rhythm of, xv
murmur, 63
ultrasound assessment of, 19
Hematologic disorders, in Down syndrome, 93-100; *see also specific disorders*
Hematopoietic progenitor cells, biphenotypic, proliferation of in leukemia, 97-98

Hepatitis B virus
  Down syndrome association with, 84-85
  vaccine, xvi, 89-90
Hepatosplenomegaly, transient leukemia
    associated with, 99
H influenza B
  bacterial tracheitis due to, 87
  vaccine, 89-90
Hippocampus, reduction in size of, 169
Hips
  dislocated, xv, 57, 118-120, 124
  fusion of, 120
  orthopedic problems in, 112, 118-119,
    124
  subluxed, 118
HMG14 gene, on chromosome 21, 8-9
Hormones, in attempted modification of
    Down syndrome phenotype, 90
Human chorionic gonadotropin (hCG), as
    biochemical screening parameter for
    Down syndrome, 14-19
Humeral length, shortened, ultrasound
    assessment of, 19
Humoral immunity, defective, in Down
    syndrome, 85, 90, 100
Hunter–Hurler syndrome, otitis media
    associated with, 88
Huntington's chorea, dementia associated
    with, 174
Hydrops fetalis, transient leukemia
    associated with, 99
Hyperactivity, incidence of, 58
Hypercarbia, due to untreated sleep apnea,
    132
Hyperechogenic bowel, ultrasound
    assessment of, 19
Hyperopia, incidence of, 149-150
Hyperthyroidism, incidence of, 72-74
Hypopharyngeal collapse, during
    inspiration, 131
Hypopharynx, structural abnormalities in, 56
Hypoplastic left heart syndrome, incidence
    of, 61
Hypothyroidism, incidence of, 56, 71-74
Hypotonia
  in alveolar hypoventilation, 65
  on chromosome 21 phenotypic map of
    Down syndrome, 7
  incidence of, xi, 105, 107
Hypoxemia, incidence of, 65, 132

IFNAR gene, on chromosome 21, 9

IFNBR gene, on chromosome 21, 9
IFNGR2 gene, on chromosome 21, 9
IFNRA/B genes, on chromosome 21, 8
IFNTI gene, on chromosome 21, 9
IGF1. See Insulin-like growth factor 1
Iliotibial band, release of, 116
Immune response, defective, in Down
    syndrome, 10, 85-86, 89-90, 100
Immunoglobulins, depressed synthesis of,
    86, 90
Impulsivity, excessive, 58
Independent living counseling, in young
    adults, xvi
Individuals with Disabilities Education Act,
    transition planning addressed in, 158
Infantile spasms, incidence of, 56, 105, 107
Infection, increased susceptibility to, x, xi,
    10, 49, 53, 56, 61, 83-91, 99; see also
    specific infections
Influenza immunization, need for, xvi, 90-91
Infrapatellar tendon, lateral insertion of, 115
Insulin-like growth factor 1 (IGF1), growth
    hormone in stimulation of production
    of, 79-80
Intelligence quotient (IQ), range of in Down
    syndrome, 104
Intrauterine diagnosis, of Down syndrome, xi
IQ. See Intelligence quotient
Iris, characteristics of, 147-148

Joint laxity, on chromosome 21 phenotypic
    map of Down syndrome, 7

"Kalmuc idiocy", historical reporting of, 3
Karyotyping, chromosomal, in neonatal and
    early infancy, xv,
Keratoconus, incidence of, 56, 147-148,
    152-154
Klinefelter's syndrome
  Down syndrome associated with, 107
  extra sex chromosome in, 23
Knee. See Patella

Language, development of in Down
    syndrome children, xi, xvi, 39-50
Launching, as concept in process of
    becoming independent, 159
L-Dopa, effect of on plasma growth
    hormone concentrations, 79-80

Learning, in Down syndrome, characterization of, 37-38
Leber amaurosis, keratoconus associated with, 152
Left heart obstructive diseases, incidence of, 61-62
Leiter International Performance Scale, in assessment of functioning levels, 171
Leukemia, increased susceptibility to, x, xi, 10, 87, 93-100
(LFA-1B) gene, on chromosome 21, 9
LH. See Luteinizing hormone
Lifespan, in Down syndrome, xi
Ligaments, lax, 114
Locus coeruleus, norepinephrine levels and neuronal numbers in, 176
Luteinizing hormone (LH), in gonadal function, 75-76

MacArthur Communicative Development Inventory, in investigation of early vocabulary development, 42-43, 45, 50
Magnetic resonance imaging (MRI), Alzheimer's dementia assessed via, 175
Mandibulofacial dysostosis, otitis media associated with, 88
Maternal age, advanced, increased risk of trisomies correlated with, 3, 14, 25-26, 28-30, 32
Maxilla, underdevelopment of, 87, 90, 135, 139-140, 143
Medial heel wedge, in shoes, 113
Medical care, meeting needs for, ix, 53-58, 164-165
Medical-surgical intervention, timetable for, in Down syndrome, xv, xvi
Megakaryoblasts, in leukemia, 96-100
Megakaryocytes, in leukemia, 98-99
Megavitamin therapy, in attempted modification of Down syndrome phenotype, 88-90
Meiosis, maternal, in origin of trisomies, 28-30, 32
Menarche, in Down syndrome women, 76-77, 160
Mental health, assessment of, 163-164
Mental retardation
    chromosome 21 genes for, 8, 10

on chromosome 21 phenotypic map of Down syndrome, 7
Mental status evaluation, in assessment of dementia, 171-173
Metabolic errors, dementia associated with, 174
Metatarsus primus varus, incidence of, 112-113
Microcephaly, causes for, 103-104
Micrognathia, otitis media associated with, 88, 130
Middle ear
    congenital anomalies of, 87
    effusion of, 127-129
    fluid accumulation in, 56
Midfacial hypoplasia, underdevelopment of, 65, 87, 130, 135, 139
Minerals, in attempted modification of Down syndrome phenotype, 88, 90
Minnesota Developmental Programming System's Behavior Scales, in assessment of social and behavioral functioning, 172
Miscarriage, trisomies associated with, 23
Mitral regurgitation, from left ventricle into left atrium, 62
Mitral valve prolapse, incidence of, 68
MMR vaccine, need for, 89-90
Mohr syndrome, otitis media associated with, 88
Monosomy, deleterious effects of, 10
Motor apraxia, assessment for, 173
Motor competence, decline in, 104
MRI. See Magnetic resonance imaging
Mucopolysaccharidosis, otitis media associated with, 88
Multiple gestation pregnancy, human chorionic gonadotropin production in, 16
Muscarinic receptors, reduction of in midbrain, 104
Muscles, poor insertions into, 111
Muscular hypotonia, correlated with increased susceptibility to infection, 87, 90
MX-1/2 genes, on chromosome 21, 8-9
Myelination, of cerebral white matter, delayed, 103, 169
Myeloid system, abnormalities of, 99
Myoclonus, tonic-clonic seizures with, occurrence of, 56
Myopia, incidence of, 147, 149

Nail infections, fungal, increased
susceptibility to, 83
Nasopharyngeal infectious syndromes,
increased susceptibility to, 87
Natural killer cells, impaired reactions of,
86-87
Neck infections, increased susceptibility to,
83
Neural tube defect screening,
alpha-fetoprotein levels in, 17
Neuritic senile plaques, in Alzheimer's
disease, 176
Neurobehavioral disorders, in Down
syndrome, xi, 106-107; see also
specific disorders
Neurofibrillary tangles, in Alzheimer's
disease, 176
Neurological disorders, in Down syndrome,
49-50, 103-107; see also specific
disorders
Neurons
delayed differentiation of, 169
membrane hyperexcitability in, 170
migration of, 103, 107
Neuropathological development, delayed
and abnormal, 168-170
Neurotransmitter deficits
in Alzheimer's disease, 176-177
disturbances of, 104
Neutrophil chemotaxis, defective, 86
Nondisjunction, in human autosomal
trisomies, 23-30
Norepinephrine levels, in Down syndrome
brains, 176
Nose, abnormalities of, 135
Nuchal fold, thickened, ultrasound
assessment of, 19
Nursing bottle caries, incidence of, 141
Nutritional concerns, of infants and children,
xvi, 57
Nystagmus, incidence of, 56, 147, 150-151

Obesity
alveolar hypoventilation and, 65
orthopedic problems and, 114
Occipital ascent, steep, 103
Occipital-frontal diameter, shortened, 103
Occupational therapy, for preschoolers, xvi
Ocular disorders, in Down syndrome, 56,
147-154; see also specific disorders
Oculoauriculovertebral dysplasia, otitis
media associated with, 88

Oculomotor imbalance, incidence of,
149-151, 153
Oculomotor programming, immaturity of, in
hypotonia, 105
Office orthopedic examination, in
preschoolers, xv
Oncologic disorders, in Down syndrome,
93-100; see also specific disorders
Operculum, hypoplasia of, 103
Oral considerations, in Down syndrome, xi,
135-145; see also specific
considerations
Orofacial-digital syndrome, otitis media
associated with, 88
Orthodontic manipulations, in attempted
modification of Down syndrome
phenotype, 88
Orthopedic disorders, in Down syndrome,
xi, 111-125; see also specific
disorders
Ossicular fixation, in augmentation of
conductive hearing loss, 128
Osteoarthritis, spinal, 121
Osteopetrosis, otitis media associated with,
88
Otitis media
increased susceptibility to, 83, 87
influenza immunization against, 90
recurrent, 127-130
Otolaryngologic manifestations, of Down
syndrome, xi, 127; see also specific
manifestations
Otoscopy, microscopic, in preschoolers, xvi
Outgoing nature, strong sense of, as
neurobehavioral aspect of Down
syndrome, 106
Ovulation, in Down syndrome women, 160

PAIS gene, on chromosome 21, 9
Palate, abnormalities of, 130, 139-140
Palpebral fissures, incidence of, 147-148
Parent report measures, of early vocabulary
production, 43
Parents
initial counseling of, 53-54
separation from, counseling of young
adults for, xvi
Parrot disease, otitis media associated with,
88
Partial complex seizures, incidence of, 56
Patau syndrome, otitis media associated
with, 88

Patella
    dislocated, 112, 115-118, 124-125
    fusion of, 117
    orthopedic problems in, 112, 114-118,
        124-125
    reconstruction of, 116
    subluxed, 114-117
Patellofemoral reconstruction, surgical, 116
Patent ductus arteriosus, assessment of, 55
Penile size, in Down syndrome men, 75, 161
Periodontal disease
    defective neutrophil chemotaxis
        correlated with, 86
    incidence of, 140, 142, 144
Periodontal therapy, in preschoolers, xv
Persistent ductus arteriosus, incidence of, 62
Personality changes, in adults, xvi
Personality characteristics, of individuals
        with Down syndrome, 58-59, 106,
        111
PET. See Positron emission tomography
PFKL gene, on chromosome 21, 8-9
PGFT gene, on chromosome 21, 9
pH abnormalities, intracellular, 79
Pharyngeal wall, lateral, medialization of,
        131
Phenotype, Down syndrome
    attempted modification of, 88, 90
    map of, 5-8
    neurological, 104
Physical therapy
    for orthopedic problems, 111, 116, 124
    for preschoolers, xvi
Phytohemagglutinin, decreased T-cell
        proliferative response of to, 86
Pick's disease, dementia associated with, 174
Pierre Robin syndrome, otitis media
        associated with, 88
Pinna, in neonates, 127
Plastic surgery, in modification of Down
        syndrome phenotype, xi, 88
Platelets, in neonates, 99
Pneumo-otoscopic examination, of middle
        ear, 129-130
Pneumococcal vaccine, multi-valent, need
        for, xvi, 89-90
Pneumonia, increased susceptibility to, 83
Poliovirus vaccine, need for, 89
Polycythemia, in neonates, prevalence of, 99
Positron emission tomography (PET),
        Alzheimer's dementia assessed via,
        175

Potassium-transport abnormalities, in Down
        syndrome cells, 79
Pregnancy
    human chorionic gonadotropin
        production in, 16
    previous Down syndrome, 14
Pregnancy specific (SP1) glycoprotein, as
        biochemical screening parameter for
        Down syndrome, 16-17, 19
Prenatal diagnosis
    biochemical screening parameters and,
        14-18
    ultrasound screening and, 19
Presbycusis, pathology of, 128
Preventive behavior checklist, for
        preschoolers and school age children,
        xvi
Prevocational adjustment counseling, for
        young adults, xvi
PRGS gene, on chromosome 21, 9
Prognathism, facial, 135, 142-143
Pronation, of feet, 113
Property destruction, by school age children,
        xvi
Pseudo-dementia, with loss of adaptive
        capacities, 105
Psychiatric disorders, incidence of, 57-58,
        106
Psychoeducational evaluation, in school age
        children, xvi
Psychometric testing, in assessment of
        dementia, 171-173
Psychomotor retardation, Down syndrome
        associated with, 10
Puberty, in Down syndrome adolescents, 75,
        160-161
Pulmonary artery hypertension
    congenital heart defects and, 55, 64-65
    sleep apnea and, 57, 132
Pulmonary hypoplasia, incidence of, 64
Pulmonary vascular obstructive disease,
        incidence of, 63-64, 67-68
Pyle disease, otitis media associated with,
        88
Pyloric stenosis, incidence of, 55

Q-angle, increased, 115
Q-banded chromosome 21
    heteromorphisms, analysis of, 25, 27

Racial factors, in human chorionic
        gonadotropin production, 15-16

Red reflex test, for eye evaluation, xv, 152-153
Reefing, medial capsular, 116
Refractive errors, incidence of, 56, 147-154
"Relaxed selection" hypothesis, in explanation of correlation of increased maternal age with Down syndrome incidence, 28-29
Repetitive behaviors, incidence of, 58
Respiratory infections, increased susceptibility to, 83, 90
Retinacular release, lateral, 116
Retinal vasculature, increased, incidence of, 147
Ridiculous, strong sense of, as neurobehavioral aspect of Down syndrome, 106
RNA 4 gene, on chromosome 21, 9
Robertsonian translocation, percentage of risk for, 14

S100β gene, on chromosome 21, 8-9
Salivary flow, reduced, 135
Salter pelvic osteotomy, for dislocated hip, 119
Scaphoid pads, in shoes, 113
Schizophrenic disorders, incidence of, 106, 163
Schooling, in Down syndrome, xvi, 58
Sclerotic panencephalitis, subacute, dementia associated with, 173
Scoliosis, incidence of, 120-121, 124
Screening, for Down syndrome, 13-19
Sealants, dental, xv, 142-143, 145
Seasonality, of birth, infection occurrence correlated with, 85
Second trimester screening, for Down syndrome, 18-19
Seizure disorders, incidence of, xi, xvi, 55-56, 105, 107, 173
Selenium, supplementation with, impact of on infection and immunity, 89-90
Self care, in preschoolers, xvi
Self injurious behavior, in school age children, xvi
Sensorineural hearing loss, incidence of, 128-130
Serotonergic neurotransmitter systems, disturbances of, 104
Sexuality, in Down syndrome, xii, xvi, 76-77, 160-163, 165

Shoes, appropriate, for orthopedic problems, 112-114, 124
"Sicca cell therapy", in modification of Down syndrome phenotype, 88-89
Sign vocabulary, acquisition of, 45-50
Sinus infections, increased susceptibility to, 83, 87, 90
Sitting postures, bizarre, 118, 124
Skeletal disorders, in Down syndrome, 57, 111-125; see also specific disorders
Skin infections, fungal, increased susceptibility to, 83
Skull malformations, correlated with increased susceptibility to infection, 87
Sleep
  patterns of in preschoolers, xvi
  plasma growth hormone concentrations and, 79-80
Sleep apnea, obstructive, 57-58, 64-66, 127, 130-132
Slosson Intelligence Test for Children and Adults, in assessment of functioning levels, 171
Snellen letter chart, in assessment of visual function, 153
Snoring, in obstructive sleep apnea, 57, 65-66, 131
Social acceptance, of Down syndrome patients, ix
Social competence, strong sense of, as neurobehavioral aspect of Down syndrome, 106
SOD1 gene, on chromosome 21, 6-9
Sodium-transport abnormalities, in Down syndrome cells, 79
Somatomedin. See Insulin-like growth factor 1
Somatostatin levels, in Down syndrome brains, 177
Southern blot dosage analysis, in estimation of gene copy number, 6
SP1 glycoprotein. See Pregnancy specific glycoprotein
Speech, development of in children with Down syndrome, xvi, 39-50
Spine
  irreversible damage to, 57
  orthopedic problems in, 112, 120-125
  subluxation of, xvi
Splaying, of foot, 112
Stance, bizarre, 112

Stanford-Binet Intelligence Scale, in assessment of functioning levels, 171
*Staphylococcus aureus*, in bacterial tracheitis, 87
*Staphylococcus epidermidis*, in blepharitis, 148
Stenosis, external auditory canal, 127-129
Stools, absence of, in neonatal and early infancy, xv
Strabismus, incidence of, 56, 147, 149-151, 153-154
*Streptococcus* spp., bacterial tracheitis due to, 87
Stubbornness, as behavior, 58, 106
Sub-talar fusion, for flat feet, 114
Superior temporal gyrus, hypoplasia of, 103
Superoxide dismutase, excess, cellular, correlated with increased susceptibility to infection, 87
Support groups, parental, in neonatal and early infancy, xv
Surgical intervention
    against heart disease, 67
    against orthopedic disorders, 111-124
    against otitis media, 129-130
Synaptic density, abnormalities in, 103
Synaptogenesis, postnatal, defect in, 103, 107

T3 screening
    in adults, xvi, 74
    in children, 72, 74
    in neonatal and early infancy, xv, 56
    in preschoolers, xvi
T4 screening
    in adults, 73
    in children, 72-73
    in neonatal and early infancy, xv, 56
    in preschoolers, xvi,
T-cells
    circulating, reduction in, 86
    decreased proliferative response of, 86
    in otitis media, 128
Teeth
    abnormalities of, 135-145
    general sequence of eruption of, 139
Teller Acuity Cards,in assessment of visual function, 153
Temporal bone, pathology of, 128
Temporal lobe, dysmaturity in, 104

Tendons, insertion of, 111, 115-116
Tensor veli palatini muscle
    abnormalities of, 56
    function of, 128
Testicular volume, in sexually mature men, 75, 161
Testosterone, in gonadal function, 75-76
Tetralogy of Fallot, incidence of, 55, 63
Thalamus, echogenicity of, 103
T-helper cells, decreased numbers of, 86
Thrombocytopenia, in leukemia onset, 96
Thyroid disease, incidence of, xi, 56, 71-74, 87
Thyroid function tests
    in adults, xvi
    in children, xvi, 72-73
    in neonatal and early infancy, xv
    in preschoolers, xvi
Tibial tubercle, lateral insertion of infrapatellar tendon into, 115
Tibia valga deformity, correction of, 116
Timetable, for medical-surgical intervention in Down syndrome, xv, xvi
Toes, gap between, on chromosome 21 phenotypic map of Down syndrome, 7
Toilet skills, in preschoolers, xvi
Tongue, abnormalities of, xvi, 130-131, 135, 139, 142
Tonic-clonic seizures, incidence of, 56
Tonsillectomy, for obstructive sleep apnea, 57, 131-132
Tonsils
    in alveolar hypoventilation, 65
    hypertrophy of, 65
Tourette's syndrome, Down syndrome associated with, 107
Tracheoesophageal fistula, incidence of, 55
Transient leukemia, in neonates, 97-99
Transitional observations, of school age children, xvi
Transition planning, for successful adjustment to adolescence and young adulthood, 157-158, 165
Treacher Collins syndrome, otitis media associated with, 88
Triple arthrodesis, for flat feet, 114
Trisomy 13-15, otitis media associated with, 88
Trisomy 18, fetal, human chorionic gonadotropin production in, 16
Trisomy 21. *See* specific aspects of Down syndrome

TSH screening
  in adults, 73-74
  in children, 72-74
  in neonatal and early infancy, xv, 56
  in preschoolers, xvi
T-suppressor cells, increased number of, 86
Tuberous sclerosis, Down syndrome
  associated with, 107
Turner syndrome
  otitis media associated with, 88
  short stature in, 79
Twin pregnancy, human chorionic
  gonadotropin production in, 16
Tympanometric examination, of middle ear,
  130

uE3. *See* Estriol, unconjugated
Ultrasound screening, for Down syndrome,
  19
Umbilical cord blood sampling, Down
  syndrome assessed via, 13
Upper airway obstruction, incidence of, 57,
  65
Uvula, removal of, for obstructive sleep
  apnea, 131
Uvulopharyngopalatoplasty, against
  obstructive sleep apnea, 132

Vaccinations, necessity for, xvi, 89-91
Varus derotation osteotomy, for dislocated
  hip, 118
Vastus medialis, advancement of, 116

Ventilation tubes, against middle ear
  effusion, 129-130
Ventral septal defect, incidence of, 55,
  61-64, 66-67
Vineland Adaptive Behavior Scales, in
  assessment of social and behavioral
  functioning, 172
Viral antigens, defective cellular responses
  to, 86
Viral central nervous system infection,
  dementia associated with, 173
Visual function, assessment of, 153
Visual impairment, incidence of, 56
Visual sequential memory, deficits in, 104,
  107
Vitamins
  dementia associated with deficiencies of,
   173
  Down syndrome phenotype modification
   via, 88-90
Vocabulary acquisition, strategies for, 41-48
Vocational training, strategies for, 58-59
Vomiting, in neonatal and early infancy, xv

Walking ability, related to foot deformities,
  112
Weight reduction, against flat feet, 114
White blood cell counts, lowered, 86
Williams syndrome, language skills in, 48

Zinc, supplementation with, impact of on
  infection and immunity, 89